International Perspectives on Aging

Volume 12

Series Editors

Jason L. Powell
University of Lancashire, Manchester, United Kingdom

Sheying Chen
Pace University, New York, New York, USA

The study of aging is continuing to increase rapidly across multiple disciplines. This wide-ranging series on International Perspectives on Aging provides readers with much-needed comprehensive texts and critical perspectives on the latest research, policy, and practical developments. Both aging and globalization have become a reality of our times, yet a systematic effort of a global magnitude to address aging is yet to be seen. The series bridges the gaps in the literature and provides cutting-edge debate on new and traditional areas of comparative aging, all from an international perspective. More specifically, this book series on International Perspectives on Aging puts the spotlight on international and comparative studies of aging

More information about this series at http://www.springer.com/series/8818

Suchit Arora

The Transitions of Aging

Springer

Suchit Arora
State Teachers Retirement System of Ohio
Columbus, OH, USA

Name of original series cover designer (key designer): eStudio Calamar, Berlin/Figueres

ISSN 2197-5841 ISSN 2197-585X (electronic)
International Perspectives on Aging
ISBN 978-3-319-14402-3 ISBN 978-3-319-14403-0 (eBook)
DOI 10.1007/978-3-319-14403-0

Library of Congress Control Number: 2015936331

Springer Cham Heidelberg New York Dordrecht London

Printed on acid-free paper

Springer International Publishing AG Switzerland is part of Springer Science+Business Media (www.springer.com)

To Sonia and Sage

Preface

The Transitions studies aging by analyzing the rates of non-communicable diseases across generations since the nineteenth century. It then examines the extent to which economic, epidemiologic, and political conditions in the childhood years of the generations had influenced the progress of aging. In recent decades, this issue of the role of childhood development has come to the fore as a vital one for grasping the sources of long-term wellbeing. The book, however, harnesses it mainly for examining another longstanding matter: what type of role does aging play in the long-term growth of per capita healthcare spending?

Examining the matter entails weaving together strands of research from biodemography, gerontology, health economics, economic history, epidemiology, macroeconomics, public health history, and from the modern debate on the prospects of aging. My hope is that scholars in each of the disciplines will benefit from the inter-weaving of ideas as much as I have benefitted from their work. I am not embarrassed to say that the book leaves some loose ends and unanswered questions. However, it is hard to imagine they will stay that way for long because the issue has become pressing in advanced economies, and is also critical to emerging ones, as their populations age more in the years ahead.

Without the support and guidance from Richard Steckel, Peter Howitt, and Paul Evans, advisors to my doctoral research more than a decade ago, the book may not have been possible. Based partly on the research, I was able to publish two related articles (Arora 2005, 2013), whose contents have been revised and reused in the book. It was apparent while writing the articles that the connections between the different topics involved would be clearer if they were presented together, in somewhat more detail, but doing so would also make the content book-length instead of article-length. And so it is to my wife, Sonia, and our son, Sage, I thank the most for their patience as the project ballooned.

Over the years, segments of the book were presented to various seminars at the Pan American Health Organization, where I benefited from comments by the participants, especially by Sir George Alleyne who encouraged the research very early on. As the project inched forward, I gained from discussions with Bernard Harris, Rick Steckel, Peter Lindert, John Komlos, Simon Szreter, John Murray, Werner Troesken, Hoyt Bleakley, Claude Diebolt, Oded Galor, Shankha Chakraborty, Geoff Bump, and Paul Sharp. It has benefitted from suggestions by the participants of the Beta Workshop at the University of Strasbourg, of the conference on health and economic policy at the CES/ifo Institute, and of the seminar at the University of Oregon's economics department. I am grateful to David Blau, chairman of the economics department of the Ohio State University, for arranging access to the University Libraries. Thank you, Peter Lindert, for sharing the data on wealth and income inequality during the nineteenth century. Bernard Harris, thank you for answering my many questions and for guiding me to key sources of information and research. *The Changing Body*, by Sir Roderick Floud, Robert Fogel, Bernard Harris, and Sok Chul Hong (2011), has been an inspiration and an invaluable resource.

To the series editors, Jason Powell and Sheying Chen, to Evelien Bakker and Bernadette Deelen-Mans at Springer International, thank you for the enthusiasm for the manuscript, and for all the support and patience to see it through in print. Many thanks to Nagarajan Paramasivam and to Rathika Ramkumar at SPi Global for meticulously typesetting a manuscript with a large number of figures and tables. The views expressed in the book are not of my employer, the State Teachers Retirement System of Ohio.

References

Arora, S. (2005). On epidemiologic and economic transitions: A historical view. In G. Lopez-Casasnovas, B. Rivera, L. Currias (Eds.), *Health and economic growth: Findings and policy implications* (pp. 197–238). Cambridge: MIT press.

Arora, S. (2013). Understanding aging during the epidemiologic transition. *Research in Economic History*, *29*, 1–69.

Floud, R., Fogel, R. W., Harris, B., & Hong, S. C. (2011). *The changing body, health nutrition and human development in the western world since 1700* (NBER). New York: Cambridge University Press.

Contents

List of Figures

List of Tables

Chapter 1
Aging and Non-communicable Disease

1.1 Critical to Healthcare Spending

As the populations of advanced countries age ever more in the upcoming decades, non-communicable diseases, the so-called diseases of aging such as cardiovascular dysfunctions, stroke, cancers, arthritis, Alzheimer's disease, among a host of others, are widely expected to sway healthcare spending. According to the Organization of Economic Cooperation and Development (OECD 2013), public healthcare and long-term care spending in its member countries were on average 6.2% of GDP in 2006–2010. Under various scenarios, including the ones about how aging may unfold, spending is projected to climb to 9.5–13.9% of GDP by 2060. Even in key developing nations whose populations are transitioning to the type of aging ongoing in advanced countries, and where many non-communicable diseases are surfacing more notably than before (WHO 2014), spending is set to climb from 2.5% of GDP in 2006–2010 to 5–10% in 2060.

In what is rare for so specialized an issue, the General Assembly of the United Nations recently adopted the declaration stating, "the global burden and threat of non-communicable diseases constitutes one of the major challenges for development in the twenty-first century, which undermines social and economic development throughout the world...." And, among the list of many risks, it recognized, "... maternal and child health is inextricably linked with non-communicable diseases and their risk factors, ..." (UNO 2012).

Despite widespread awareness of such diseases becoming a vital long-term issue, and a growing concern that they may stretch government healthcare budgets in the years ahead, the progress of such diseases since the distant past and their relation to healthcare spending is not well understood, especially in a country that succeeded in transitioning to an advanced economic status, and was able to do so

S. Arora, *The Transitions of Aging*, International Perspectives on Aging 12,
DOI 10.1007/978-3-319-14403-0_1

while its population was aging increasingly. The book narrows the gap by studying the matter for the United Kingdom (UK) over the last two centuries.

As the population of the UK ages further in the years ahead, the OECD projects that the public health and long-term care expenditures will grow from 7.4% of GDP in 2006–2010 to anywhere between 9.9% and 13.9% of GDP by 2060 (OECD 2013, Table 8). Where it ends up in the range depends partly on the types of aging scenarios that may play out. Informing the set of potential scenarios, in turn, is a vast body of research that has for the past few decades engaged biodemographers, health economists, economic historians, epidemiologists, gerontologists, macroeconomists, public health historians, and researchers in contemporary public health.

1.2 Grasped by Two Parameters

A quantitative tone makes it easy to see the scenarios and to grasp the role they may play in healthcare spending. Figure 1.1a sets up an example to clarify their role. On the vertical axis, expressed as a percentage of per capita income, is the per capita healthcare expense, incurred over various ages of the adult life cycle mentioned on the horizontal axis. It is an age-profile of healthcare spending. The expenditures shown in this one are the average of such age-profiles from 20 OECD countries. Meanwhile, the adult-age interval covered by the profile, age 25–30 years to 80–85 years, is that segment of the life cycle when the rates of non-communicable diseases like heart disease, cancers, stroke, arthritis, and so on, grow increasingly. The features of this type of profile form some of the main elements of how spending over time may relate to the path of non-communicable diseases and aging. It helps to outline the features before examining the scenarios such profiles help depict:

First, it is a *period*-age-profile. The 'period' part is important to bear in mind. There can be other types of age-profiles such as cohort-age-profiles, which will be discussed below. But this one refers to expenses incurred in a specific period or year—say, 2010. To keep things simple for now, say also the age groups making up the profile are the only ones spending on healthcare. The real amounts they spend per capita sums up to the spending per capita economy-wide. Thus, tagged to each such profile is a particular amount. If there were a series of annual period-profiles through 2010, they would stand for the annual time series of real per capita spending through then. Embedded in the time series would be the historical relationship of aging and healthcare spending. Likewise, a prospective series of such profiles over 2011–2060 would be the projected time series of healthcare spending. Embedded in the projections are the envisioned scenarios of aging, which partly govern how this 2010-profile may change and sway spending in the years ahead.

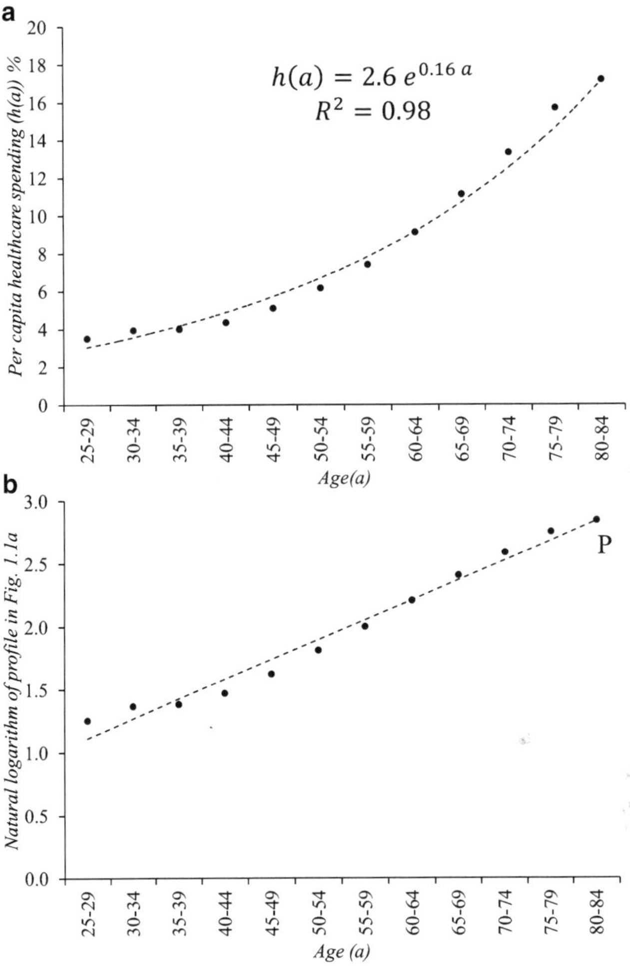

Fig. 1.1 The period-profile of per capita healthcare spending as % of GDP per capita (**a**) Exponential trend (**b**) Natural log of the profile in panel **a** (Sources: OECD (2013); author's calculations)

The upward slope of the profile is its next important feature. It is expressed quantitatively in Fig. 1.1a by an exponential curve (the broken line), which tracks the profile closely. Spending grows exponentially with age presumably because illness rates from various diseases climb with age, and perhaps because the medical treatment becomes increasingly difficult and expensive as people get older. The age-profiles of non-communicable diseases, the ones explored extensively in the book, also climb exponentially with age. This exponential path of aging was discovered long ago by Gompertz (1825, 1872), and has since been vital to research in biodemography (Makeham 1867; Olshansky and Carnes 1997). Applied to per capita spending in any year *t*, it is expressed as

$$h(a,t) = \alpha e^{\beta a}, \tag{1.1}$$

where *h*(*a*, *t*) is spending at age *a*, α is the initial state of the spending, *e* the exponential term, and β the profile's curvature or its slope. The R-square for the equation conveys that α and β—just two parameters—may help closely track the spending profiles.

The parameters make it easy to interpret the spending in the profile and enable cross-profile comparisons. For the equation estimated in Fig. 1.1a, $\alpha = 2.6$. All else equal, higher the α, more elevated the profile would be from the outset, and greater would be the spending in it. Meanwhile, the slope coefficient β conveys how fast the spending climbs. To see why, take the natural logarithm of Eq. 1.1, and then differentiate with respect to age: all else held constant, bigger the β, steeper the slope, faster the *h*(*t*, *a*) climbs. To calculate how fast for the equation in Fig. 1.1a, divide the natural logarithm of 2 (or 0.693) by the $\beta = 0.16$. It equals 4.33, which says that *h*(*a*, *t*) had doubled every 4–5 years of age. The natural logarithm of the profile shown in Fig. 1.1b expresses this very feature: expenses grow with age at a nearly constant rate, doubling every 4–5 years. The scenarios depicted below will be in straight upward-sloping lines, like the one in Fig. 1.1b, instead of the curve in Fig. 1.1a.

The parameters, moreover, clarify how spending can relate to underlying scenarios of aging. Compared to the profile P in Fig. 1.1b, a scenario where the profile has a larger initial state (α) but the same slope (β) would be a parallel line above P. At all ages, per capita expense along such a line of aging would be uniformly more than the one along P. A profile with a greater α but smaller β is one that would start above P, but at older ages will converge toward it. The spending here is larger too, but comes ever closer to the one in P at advanced ages. This shift is not parallel. The profile with smaller α and larger β starts below P, but converges toward P from below, and has less spending at each age. There can be more possibilities, but for now these three suffice for illustrating how the parameters make it easy to interpret the scenarios.

Before we delve into the scenarios, however, another vital thing to know about a profile like P is that it is made up of spending by people from many birth-cohorts. Using round numbers, the 80-year-olds in 2010 were born in 1930 (i.e. 2010 minus 80 = 1930); the 70-year-olds in 1940; the 60-year-olds in 1950; and so on until the 30-year olds, who were born in 1980. The per capita outlays at each age actually come from different birth-cohorts. It is these cohort contributions that had summed up to the economy-wide per capita spending in 2010. Now move forward a decade to 2020. The cohorts will have aged ten years. Some of the 70-year-olds in the 2010-profile, for instance, will be the 80-year olds in the 2020-profile. Meanwhile, a new cohort (the 1990 cohort) will have joined the mix, contributing for the first time at age 30 to the 2020-profile. Along with other proximate cohorts, it will then contribute to the period-profiles in several ensuing decades until it is 80 years old in 2070.

This cohort-composition or cohort mix clarifies an important issue: P can shift if the cohorts contributing to it exhibit some type of change. The change can appear as early as age 30 since each new cohort begins contributing when it is 30 years old. If a series of cohorts exhibit change in a particular direction, they can shift many period-profiles in that direction, helping form a trend in the period-profiles, and thus forming a trend in the time series of what the profiles stand for. To see this better, say the period-profile were measuring the rates of non-communicable diseases at each age. They would sum up to the aggregate rate of non-communicable diseases in the year. If those age-specific rates were to shrink in several successive cohorts, all else equal, the series of period-profiles may keep shifting down. This would be depicted in the same way as a series of lines increasingly below P in Fig. 1.1b. Since each period-profile stands for an aggregate rate of non-communicable diseases, the series of downward shifts would stand for a falling trend in the aggregate rate.

1.3 That Clarify Some Scenarios

With those features of the profiles in mind, imagine an aging scenario in which the period-profile is said to be static over the next several decades: the profile in 2010 is the same as the one in 2020, 2030, and so on through 2060. This is equivalent to saying that *all* cohorts contributing to those profiles will stay more or less identical. They will all trace some universal profile like P, which essentially functions like a fixed set of place-holders at each age. Since the generations neither differ in their spending, nor in their underlying physiological condition that could tell on the spending, each one would occupy more or less the same places as it passes through. For example, each generation may have identical illness and death rates of non-communicable diseases, and the total rate of such diseases stays the same. All else constant, the logical result of such a static or a stationary spending profile should be that spending is impervious to aging *across* generations, mainly because the aging of any one generation is not different from that of the others.

Now consider Fig. 1.2 that lists what are said to be some of the key factors boosting per capita healthcare spending. On the left side are the two aging-related scenarios: (1) the average age in 2060 will be greater; and (2) the health status of the middle-to-older aged in 2060 may improve. They are stylized in Fig. 1.3a–b, with spending per capita $h(a)$ on the vertical axis, and age (a) on the horizontal one.

In Fig. 1.3a, as the average age climbs in the years leading up to 2060, spending slides up along the profile. This scenario of aging clearly presumes a stationary profile over the next several decades. The sliding up along a given profile is called the pure aging effect, or, "... the mechanical effect of population aging on expenditures, ..., assuming that the age-profile of expenditures remains constant over time..." (OECD 2013, p. 15).

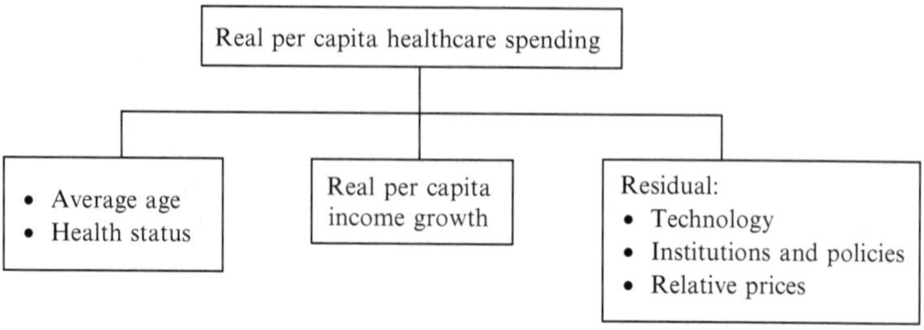

Fig. 1.2 Main influences on real per capita healthcare spending (Source: OECD (2013))

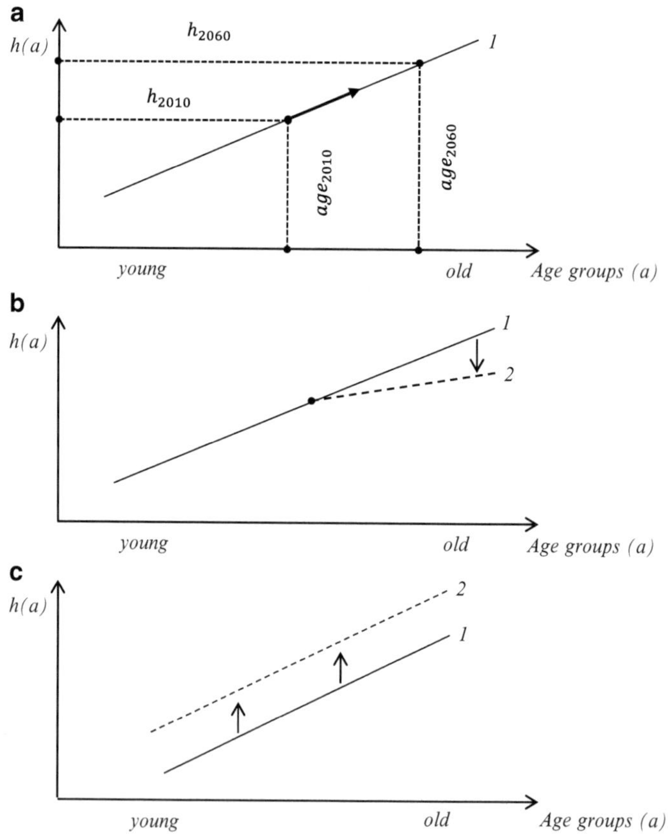

Fig. 1.3 Stylized scenarios for the age-profile of healthcare spending *h(a)* (**a**) Pure aging effect (**b**) Aging effect adjusted for death-related costs and healthy longevity (**c**) Non-aging drivers (Sources: OECD (2013, Fig. 5); author's interpretation)

However, assessing the role of aging in that way blocks a critical issue from telling on the outcome. The method starts with a given static spending profile. It alters the age-composition, envisioning a greater percentage of people at older ages than at younger ones. It then asks what would be the spending as that happens. But a problem comes up at this point. The new age-composition, itself, may mean that the spending profile may not remain static. For the average age to have grown, the new age-composition must have a larger percentage of older people. Being older-aged, they may have greater rates of non-communicable diseases, and may need to spend more than what the baseline spending-profile indicates. The reverse could happen if the rates for some reason were less. The very purpose of altering the age-composition, of saying that there will be more older people than today, is to find out what would happen if the age-profile of spending itself is steeper (smaller α, larger β) or is flatter (larger α, smaller β) than the baseline static profile (unchanged α and β). Holding the profile static or stationary could then misread the ramifications of aging. The issue becomes acute when the stationary profile is used to build in the scenario of healthy longevity or healthy aging. In this type of scenario, also called the 'compression' of deaths and illness (Fries 1980, 1989, 2005), older people are seen as healthier than they used to be. As deaths are delayed to ever older ages, people become ill only in the very waning years of life, just before some limiting age at which death is more or less inevitable. So illness becomes increasingly 'compressed' against that age limit. A smaller percentage of the growing lifespan will have been spent in illness, requiring less healthcare expense than before. Consequently, the projected static profile of spending bends at middle-to-older ages, making its last few age-segments flatter than the rest (Fig. 1.3b). The opposite scenario, called the expansion—of more time spent in illness, requiring more healthcare spending—would be depicted by an upward tilt in the last segment, making it steeper than the rest of the age-profile.

However, this way of depicting healthy aging has a stark asymmetry in the underpinnings: cohorts of older people continually fare better, but the youth just stay the same. The youth show no sign of what their state might be at older ages. The spending profile then acts accordingly: spending rates for the young-age segment of the profile stay unchanged, but the slower spending built-in for the healthier older-aged people makes the overall profile flatter than it would otherwise be. Consequently, the scenario of healthy aging delivers slower spending than the baseline static case.

The three potential scenarios in Fig. 1.3a, b—static profile, semi-static profile associated with compression, and the one associated with expansion—envision specific underpinnings of aging, and of the accompanying rates of non-communicable diseases. Which one of them is likelier than the others? We do not know. But what we can do is study the scenarios that played out in the past: when did the expansion occur or when did the compression take hold, and how did one transition to the other? *The Transitions* explores just this matter since the nineteenth century.

It finds that over the long horizon: (1) the profiles shift sizably, undermining the premise of a static profile; (2) in episodes of progress or of regress, period- as well as cohort-profiles shift since early adult age; (3) flatter slopes appear in profiles when a series of cohorts were faring worse, and faring worse since early in their adult life cycles.

Attached to the relatively flat segment of the profile in Fig. 1.3b, then, an unchanging segment for the youth seems out of place: they ought to have been faring worse, and the spending likely would have to respond. Moreover, the evidence conveys that instead of being the good scenario of healthy aging, a flatter profile in non-communicable diseases stands for the expansion of deaths, and likely for the expansion of illnesses leading up to the deaths. The book explores the circumstances when slower spending may accompany this type of aging; they do not seem benign.

To crystallize the matter in quantitative terms, the core issue in the prospects of aging, and of the associated per capita healthcare spending, could be stated as: if the profile P in Fig. 1.1b shifts in the future, what type of a shift is it? Is the new profile above or below P? Is it steeper or flatter than P? Since an underpinning of P is the profile of non-communicable diseases of aging, how have the αs and the βs of the aging-profiles evolved over the long term? How may they relate to the αs and the βs of the spending profile like P?

The basic elements of the exploration develop in Chap. 3. It estimates the parameters the profiles of non-communicable diseases in 62 cohorts. A cohort is defined as people born during particular decades since the nineteenth century. In contrast to the period-profile, a cohort-profile traces the rates a particular cohort had faced as it aged over the decades say, every ten years, as it progressed from 30 years of age through age 80. The cohort-profiles were unique. They had shifted down over the long term. As their α fell and β grew, the profiles became progressively steeper. This is the good outcome. The aging cohorts had contributed to the various period-profiles along the way, as described earlier. The period-profiles too then shifted down and became steeper, forming a downward trend in the aggregate rates of non-communicable diseases.

However, there is one prolonged episode in the nineteenth century when the profiles had shifted up for a series of cohorts. This is something rare in the history of aging. Those cohorts' α grew and β fell—the bad outcome—as did the parameters of the period-profiles the cohorts shaped over about 50 years, leading to an upward trend in the aggregate rates.

So, instead of a single static profile, there were several unique profiles, one for each cohort, and they look like the profiles in Fig. 1.4. In episodes of progress, the new profiles started at a lower point, and were steeper than previous ones (Fig. 1.4a): rates of non-communicable diseases in the newer profiles fall at each age, and deaths are delayed to progressively older ages—the compression. The opposite, or the expansion, occurs in episodes of regress (Fig. 1.4b). Compression is associated with more aging, expansion with less. Notice that in both types of episodes, the worst profile, the one with highest rates, is the one on the top. Its slope is less than the rest. The smallest β pairs with the least aging potential, which likely is not a sign of healthy aging. All told, compression is desirable,

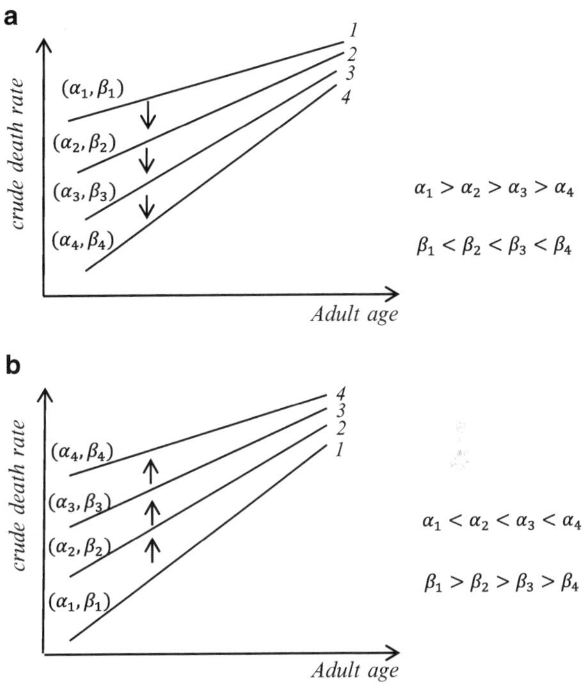

Fig. 1.4 Stylized scenarios of shifts in the age-profile of non-communicable diseases (**a**) Downward shifts (compression) (**b**) Upward shifts (expansion)

expansion is not. The next section places these scenarios of aging in the arena of healthcare spending.

1.4 And the Potential Choices for Policy

The scenarios figure in a framework of accounting that offers various options for healthcare policy. The factors influencing healthcare spending shown in Fig. 1.2 are usually analyzed in the equation:

$$g_n = \mu_y g_y + \mu_a g_a + g_r, \tag{1.2}$$

where, g_h is the growth of real per capita healthcare spending, g_y real per capita income growth, g_a is the growth of aging, and g_r growth of the residual factors. The coefficient μ_y is the income elasticity, or the percentage change in spending for a percentage change in real per capita income; and μ_a is the aging elasticity, or the percentage response of spending to a percentage change in aging. The residuals, however, are not so well defined. They are said to be some mix of technology influencing the medical sector's productivity (which is notoriously difficult to isolate), the institutional practices and policies followed by the sector (e.g. access to insurance; modes of funding and of delivery), and change in relative prices (e.g. wages of healthcare workers relative to others; Baumol 1967, 1988). The residual factors make up the supply of healthcare; real per capita income and aging, the demand (e.g. see Olliveira Martins and de la Maisonneuve 2006; OECD 2013).

Moreover, Eq. 1.2 is not a setup of causality, but of accounting and analysis, much like the Solow growth model (Solow 1956). Scholars usually estimate μ_y and μ_a in experiments and studies, and then insert the estimated values in Eq. 1.2 to analyze potential outcomes, and the policy choices they offer.

The discussion in the previous section offers a way of analyzing the roles of μ_a and g_a in this framework. Before seeing how, the first issue is why does it matter? It matters because the mix of factors that make up the residual play a pivotal role in policy. Notwithstanding the difficulties of measurement, the supply side, as embodied by the residual, is usually the target for squeezing out inefficiencies in the healthcare sector. But how big is the residual in the scheme of things? The heft of the residual in Eq. 1.2 hinges on the correct assessment of the roles of income growth and aging:

$$g_r = g_n - \mu_y g_y - \mu_a g_a, \tag{1.3}$$

which means that if the other two factors are not accounted for properly, the scope for maneuver offered by the residual factors can be misgauged.

By assuming average age as the measure of aging g_a, an implicit value of 1 for μ_a, and of 0.8 for the income elasticity μ_y, the OECD (2013), for instance, calculates that of the 4.6% g_h over 1995–2005, just 0.2% is an 'age-effect' and 1.5% 'income effect.' The rest 2.9% is the residual's growth g_r. It then uses those short-term estimates to frame scenarios over the next six decades, inferring that "pure demographics and income effects will play only a minor role in the projected increase of public health and long-term care expenditures." This offers a very wide scope for what policy could accomplish by tweaking the various factors in the residual. It also leaves the impression that trimming the residual can be a free lunch.

Before examining the difficulties such tweaking could involve, however, it is important to see how each element in Eq. 1.2 can change the spending profile. In Fig. 1.2c, the parallel shift upward in the profile is from $\mu_y\, g_y$ and from the residual g_r. By OECD's estimates, the shift is less from $\mu_y\, g_y$ than from g_r. As real per capita income grows, people at each age just spend more on healthcare; proportionately if $\mu_y = 1$, more than proportionately if $\mu_y > 1$. In one of the baseline scenarios, for example, the OECD uses $\mu_y = 0.8$, so the parallel upward shift is less than proportionate: 1% increase in real per capita income by, say, 2020, boosts per capita healthcare spending by 0.8% at each age in the 2020-profile. More real per capita income thus uniformly imbues the entire profile with more spending. Meanwhile, as the residual grows, healthcare costs climb at each age for various reasons, also resulting in a parallel upward shift. The residual and real income, therefore, are the only reasons for the shift. Aging is not the shift factor. It just moves the spending along the static profile.

That the residual boosts g_h the most leaves the impression that it can be squeezed to slow spending under the rubric of cost containment, which has been in vogue, often contentiously, since the 1980s and 1990s (Schwartz 1987; Baumol 1988; Culyer 1989; Weisbrod 1991). However, that rests on the premise that α and β of the profiles stay the same across decades. The historical underpinnings of non-communicable disease profiles, however, do not support such a premise. This is why studying the role of μ_a and g_a matters.

It matters also because a growing body of research in health economics has adopted the assumption of a static aging-profile to assess the role of aging. And by adopting that assumption, a burgeoning literature has been increasingly skeptical that aging sways healthcare spending by much, even calling it a 'red herring,' one that has somehow obscured the 'truth' as well as the 'right' policies for reining in the healthcare sector's growth (Zweifel et al. 1999; Werblow et al. 2007; Reinhardt 2003; Breyer et al. 2011).

This static way of assessing the issue is not new, however. It appears to have a long history, going back to at least Abel-Smith and Titmuss (1956), who first surmised the costs of the National Health Service two decades out, though, unlike the recent literature, the authors were acutely aware of the fragility of the premise of a static profile. More recently, however, few scholars, notably Fuchs (1990) and Newhouse (1992), have raised doubt about the premise:

> Under the assumption that the cross-sectional age-spending relation holds constant over time, the effect of the change in the age distribution of the population is estimated by applying the cross-sectional data to the change in the age distribution......Although this method is widely used...it is problematic...[as] the age-expenditure pattern may change over time.... When age-specific death rates fall over time, there are fewer people in the last year of life at any age; thus their expenditures may be less than those predicted from a previous age-spending pattern. (Fuchs 1990, p. 536).

Here "the cross-sectional age-spending relation holds constant over time" refers to a given static period-profile of spending. "When age-specific death rates fall over time, there are fewer people in the last year of life at any age" expresses the compression of deaths like the ones shown in Fig. 1.4a; and, "thus their expenditures may be less than those predicted from a previous age-spending pattern" conveys that the period-profile of spending would shift downward as deaths compress. Chapters 9 and 10 explore under what additional conditions other than the compression the "expenditures may be less." Meanwhile, citing Fuchs (1990), Newhouse (1992) states:

> ...as age-specific mortality rates fall, fewer people at each age are close to death. This would cause the method just described [static profile] to understate the effect of aging on cost.... Falling age-specific mortality rates may increase the propensity to intervene medically at any age, which provides an alternate reason why an aging population may consume more medical care. (Newhouse 1992, p. 6).

Here "age-specific mortality rates fall and fewer people at each age are close to death" is akin to deaths being delayed, or to the compression. The "increase in propensity to intervene medically at any age" seems to be about the medical spending profile's new slope as the compression occurs. While both Fuchs (1990) and Newhouse (1992) observe the importance of this issue, neither one offered a way to assess it, a gap this book narrows.

Drawing from the evidence on shifting profiles, the book questions the burgeoning view that aging may not play a major role in per capita healthcare spending. The measure of aging it uses is the series of βs across the profiles. The growth of aging is the growth of the βs, or $g_a = g_\beta$. As shown in Fig. 1.4a, shrinking rates of non-communicable disease in the profiles associate with more aging, steeper slopes, growing β, and thus positive g_β. Aging is thus defined as the shift in the profiles since early in adulthood. This relaxes the assumption of a static profile of non-communicable diseases as an underpinning of the healthcare spending profile. Since now the underpinnings can shift, so can the spending profile.

The βs of the series of period-profiles of non-communicable diseases, it turns out, relate closely to per capita healthcare spending. Over 1960–1984, their growth accounts for about 25% of g_h; over 1985–2000, nearly 40%. Thus, the role aging can play may have been bigger than what could be found by keeping the profile static, which assumes away modern aging's main thrust, the changing βs.

When the influence of aging is assessed by letting the profiles shift, the policy implication attached to the residual becomes suspect, even socially sub-optimal: cost containment may not be a free lunch if the costs are the reason why non-communicable diseases were shrinking to foster aging. In other words, if the costs co-vary positively with aging, then reining them in may subdue aging itself. Besides being adverse to wellbeing, it could hold back optimal allocation of resources to the healthcare sector that may otherwise grow in line with the high value people place on its growth; hence, the sub-optimality.

Chapter 9 explores this matter in some detail. Chapter 10 goes on to analyze why aging could have related to per capita healthcare spending over the long term. It studies analytically the transitions of aging, and how the profiles of healthcare spending may evolve as the transitions occur. The spending profiles in the chapter are set up as functions of the illness- as well as aging-profiles—and thus as functions of their βs. The following conditions seem necessary for per capita healthcare spending to have grown since the nineteenth century: (1) compression in the aging profiles co-vary negatively with the shifts in the reported-illness profiles; (2) norms, institutions and diagnostic technologies underpin the expansion in reported illness profiles (Johansson 1991; Weisbrod 1991; Newhouse 1992); (3) prevention of illness, with an intertemporal link, fosters more aging, but helps trim spending; and (4) for given profiles of aging and illness rates, technologies that shrink treatment costs slow spending without an adverse effect of letting illnesses go untreated. Particularly relevant is the benefit from the intertermporally-connected prevention costs as they deliver better childhood-linked aging and slower spending. Relevant to the issue are also new technologies that deliver better outcomes for the life-course component of aging at lesser unit costs than before. The analytical outcomes make explicit some of the issues in cost containment. The next section elaborates further on this matter.

1.5 But It Takes a While to See How

To reach the analysis in Chaps. 9 and 10, however, the book explores the history of modern aging, quantifying it in terms of α and β. Those 62 cohorts since the nineteenth century are examined in Chap. 3. Their profiles had swayed the period-profiles 1850 through 2000. The cohort-profiles have their own αs and βs, and, as outlined earlier, the parameters of several cohorts shape the α and β of each period-profile. In the framework the book adopts, five cohorts, each at a different stage of its adult life cycle, shape each period-profile.

The evidence shows that cohort-profiles had shifted sizably as early as age 25–34. The expansion (the bad scenario) had occurred in a series of cohorts brought up during circa 1820–1850/1860. Their αs grew and βs shrank; and the profiles became increasingly flatter (like Fig. 1.4b). Then, the compression occurred in cohorts born during circa 1870/1980–1940s. Their αs shrank and βs grew; the profiles were increasingly steeper (like Fig. 1.4a).

To test the worth of the cohort αs and βs, Chap. 4 asks: are they capable of simulating the history of the period-profiles, and of the consequent time series of non-communicable diseases? The cohort parameters come close.

The next few steps in the book are underpinned by two findings uncovered in Chaps. 3 and 4.

The first finding that the cohort-profiles had shifted since early adult age (25–34) prompts the question: what issues could have led to the shifts as early as age 25 in the cohorts' adult life cycles? Before 25 are mainly the childhood years. This is where the link between children's development and non-communicable diseases in the adult years enters the picture. The question then becomes that of a cross-cohort inquiry: what was so different in the childhood years across the cohorts for them to have traced unique profiles over the adult years?

The second finding is that the parameters of the cohort-profiles are capable of generating simulated period-profiles, which closely approximate the actual period-profiles. It conveys: (1) the parameters of the cohorts' profiles may be used, as dependent variables, to extract the childhood-linked signals, i.e. to what extent are the parameters sensitive to the cohorts' childhood development, yielding the childhood-related signal in the parameters; (2) those childhood signals can be harnessed to generate the childhood-linked cohort-profiles; and, several such profiles, in turn, generate the childhood-implied period-profiles, as well as the associated aggregate rates of non-communicable diseases; (3) the difference between the childhood-implied period-profiles and actual ones yields the approximate life-course-related segment of the period-profiles, and of the rates of non-communicable diseases (Ben-Shlomo and Davey Smith 1991; Davey Smith et al. 2000, 2001; Ben-Shlomo and Kuh 2002; Kuh et al. 2002; Kuh and Davey Smith 2004). The exercise produces approximate contributions to aging by issues that may have influenced childhood development, and by the ones that appeared over the life-course of the cohorts. The life-course ones, by definition, are beyond the reach of the childhood influences. The result, in turn, informs the analysis in Eq. 1.2, now written as:

$$g_n = \mu_y g_y + \mu_a \left(g_\beta^C + g_\beta^L \right) + g_r. \tag{1.4}$$

Here, the measure of aging $g_a = g_\beta$ is expressed in terms of its two components: g_β^C linked to childhood, g_β^L linked to life-course; and assuming that they do not interact, $g_\beta^C + g_\beta^L = g_\beta$.

An example may clarify the relative roles of the childhood and the life-course components. Say rheumatic fever in childhood, or some debilitating childhood infection, tells on heart disease in the adult years—an intertemporal link. All else equal, preventing rheumatic fever in children may lead to fewer deaths from heart disease decades later, and thus to more aging. This would be captured by g_β^C. Besides prevention efforts in the past, effective technology for treating heart disease is another reason why deaths from it may shrink. When such technology comes along, it would expand the role of g_β^L and shrink the role of g_β^C in the heart disease death rates.

More broadly, g_β^C and g_β^L can vary depending on how children had been faring in the past, and on what types of life-course events, protections, and risks they faced after childhood. The two may also vary by non-communicable disease, since the technology for various diseases had appeared at different points in time. They may differ by the childhood proxy being used to extract the signal as well, since a given proxy for childhood development may be linked to specific adult-age diseases but not to all. Further, either component does not have to stay positive; both childhood-linked and the life-course-related outcomes can be adverse: g_β^C can turn negative, for instance, if a series of cohorts had been faring poorly as children, and g_β^L can be negative if there are diseases such as various types of cancers that may not have any childhood links whatsoever, but stubbornly defy medical therapy in the adult years. Their relative roles are explored in Chap. 8. Chapter 9 then uses g_β^C and g_β^L implied by childhood proxies along with some macro variables in the childhood years, both outlined below, to explore the options in Eq. 1.4.

The two components help make the choices for policy clearer than either the residual or the composite g_a were able to clarify in Eq. 1.2. In Eq. 1.4, healthcare policy seeking slower or more sustainable spending may find that constricting income elasticity may be outside its purview, though this cannot be ruled out for reasons discussed later. Spending cuts could slow real per capita income, but for the time being say that the long-term trend rate of economic growth is outside the policy's reach as well. Meanwhile, policy also cannot reverse the childhood-linked g_β^C, a lasting effect of the past, or of the momentum of history so to speak.

The option left on the table is the residual (g_r). It can be squeezed safely only if two conditions are satisfied: (1) its constituents are known precisely; (2) they either relate inversely to the life-course component of aging g_β^L, or are uncorrelated to it.

That is, trimming the residual makes sense only if it fosters more aging or does not damage aging. If g_r and g_β^L relate positively instead, all else equal, smaller the g_r at hand, and larger the available g_β^L it can affect, harder the options become, and likelier is the harm as well as the social sub-optimality of such policy. If g_β^L is already adverse, then squeezing g_r may only worsen a bad situation. The results indicate that g_β^C accounts for about 36% of aging and its contribution to per capita healthcare spending in the post-World War II period through 1984, and for much of the slowdown in spending over 1985–2000. I will return to this framework shortly.

Keeping those issues in Chaps. 9 and 10 in mind, the setup of Chaps. 2, 3, 4, 5, 6, 7, and 8, going backwards, may be easier to see:

Chapter 8 explores how childhood development across cohorts had influenced the rates of non-communicable diseases in their profiles. The childhood proxies used for the purpose are recommended by a body of micro-level research on the long reach of children's development on such diseases (Waaler 1984; Barker and Osmond 1986a, b; Barker et al. 1991; Barker 1992, 1994, 1998, 2004; Paneth and Susser 1995; Jousilahti et al. 2000; Barker and Lackland 2003; Fogel and Costa 1997; Steckel and Costa 1997; Costa 2002, 2003; Finch and Crimmins 2004; Crimmins and Finch 2006; Floud et al. 2011; among others). Critical to the debate on aging since the 1980s, what this research has found at the micro level, in cross-sections of middle-to-older age people, over relatively recent decades, the book extends to the aggregate-level, for adult-age *profiles* over age 25–74, across multiple cohorts, whose childhood years varied distinctly since the nineteenth century.

What made the childhood years distinct across the cohorts were the epidemiologic, economic, and political milieus, explored in Chaps. 2, 6, and 7. They convey that history of modern aging may not be grasped well without exploring fundamental economic and political change over the past two centuries. In that sense, aging may not be purely a biological issue, though biological mechanisms would be the channels through which social change could be transmitted over time, or intertemporally, to outcomes of aging (Dasgupta 1993; Fogel and Costa 1997; Floud et al. 2011).

Chapter 2 explores the epidemiologic transition (Omran 1971, 1982; Olshansky and Ault 1986). It finds that children who grew up in the first three quarters of the nineteenth century faced high rates of infectious diseases; and, by modern standards, their average heights were poor, even falling for some cohorts. Childhood infectious disease rates explored in this chapter are used as one of the childhood proxies in Chap. 8. Meanwhile, rates of non-communicable diseases, too, used to be much higher in the nineteenth century, when children's epidemiologic situation was poor, than in the twentieth, when the situation had improved drastically.

Chapter 6 explores the macroeconomic and political backdrops to the changing epidemiologic situation. It finds that since the Industrial Revolution, children by and large grew up under increasing prosperity, though often beset by volatility such as the one in the inter-war years. This aspect is gauged by the volatility-adjusted real per capita income in the childhood years.

Moreover, children in the nineteenth century received a milieu of vast inequality of income and wealth (Lindert 2000; Atkinson and Piketty 2007, 2010; Piketty 2014).

The inequality is gauged in two ways. The first one is average adult stature across the social classes. It comes from research in anthropometric history (Floud et al. 1990; 2011; Steckel 1979, 1995; Harris and Floud 1997; Harris 2001; Komlos 1993a, b; 2008). Shrinking average stature is a powerful indicator of widening inequality, and growing average stature signifies the reverse. It has one rare trait that other measures of inequality do not possess. Since stature stays unchanged after the onset of adult years, the inequality, manifested in average growth of children across the social spectrum, becomes portable, carried that way into the adult-age profiles until years later. Chapter 8 uses average stature as one of the proxies of childhood development of the cohorts to study how it swayed the αs and βs of their profiles. The other gauge of inequality is the functional distribution of national income between labor and capital in the cohorts' birth decades (Allen 2009; Piketty 2014). It is an aggregate-level measure of the trends in wealth inequality. Since the ownership of capital has historically been very concentrated, especially so in the nineteenth century and during early industrialization, inequality widens when the functional share of labor falls and that of capital climbs. Both types of inequality, extreme for children brought up through the mid- to late-nineteenth century, have narrowed over the long-term, but the second one has started widening over the past few decades.

Chapter 8 uses the macroeconomic variables alongside the childhood proxies, including rates of pregnancy complications during the cohorts' birth-decades, to extract the childhood-linked signals of aging in the cohort-profiles (Barker 1992; Leon et al. 1996, 1998; Osmani and Sen 2003). The signals thus should not be seen as some purely biological issue, for they likely reflect the effects of past socio-economic, epidemiologic, and political choices on children's development.

It is these signals that help assess the role of g_β^C, and thus of g_β^L in Eq. 1.4. It turns out, the g_β^C over the post-World War II era pertained to cohorts whose childhood years were decades earlier, going as far back as the 1880s. The signals thus typify the momentum of history. In that sense, g_β^C was likely exogenous to per capita healthcare spending after World War II. It is g_β^L that may have related closely to the concurrent healthcare costs, the ones stemming from the supply side factors making up the residual g_r. Cost containment may thus require careful scrutiny for how it may relate to g_β^L. At the same time, it may have to be careful about how children, and thus future g_β^C, may be affected. Good outcomes for g_β^C, Chap. 10 argues, tend to subdue healthcare spending until years later, something that should further the goal of containing costs though the results would not show up immediately.

Furthermore, political inequality appears to have been vital to the history of modern aging, and very likely to healthcare spending since the nineteenth century (Acemoglu and Robinson 2000, 2001, 2008). Besides being relevant to the historiography of modern aging, this issue is critical to the role income elasticity plays in Eqs. 1.2, 1.3, and 1.4.

In the first half of the nineteenth century, when the epidemiologic situation was very severe, the polity had been beset by chronic patronage, and by paralysis of reform. The patronage and the paralysis were likely fostered by the inequality in voting power (Szreter 1997, 2005). The first major reform, the Reform Act 1832, had expanded the franchise modestly; the Second Reform Act 1867 then boosted the electorate from 20% of adult males to 33%, but women could not vote. The Third Reform Act 1884 set up more uniformity than before, allowing just one Member of Parliament from most districts, boosting the representation of the urban masses who had faced the brunt of the epidemiologic situation.

This cluster of political reforms in the 1860s and the 1870s, especially the Second Reform Act 1867 and the Municipal Franchise Act 1869, appear to have undercut the paralysis at the national and the municipal levels. The waning political inequality coincided roughly with some of the more far-reaching public health reforms in the UK's history: the Sanitary Act 1866, Public Health Act 1872, Public Health Act 1875, Factory and Workshop Act 1878, Prevention of Lead Poisoning Act 1883, among a host of others (see Fig. 6.5, Chap. 6). The proximity of the two reform clusters strongly indicates that inequality of one type (political: franchise) may sway inequality of another type (of access to water, to sewerage, and the unsymmetrical exposure of the social classes to infectious diseases (Sen 1984, 1999; Deaton 2013)).

Chapter 7 then exploits the natural experiment the reforms offer for understanding the progress of aging. They partition the childhood years of the cohorts into two eras, pre-reform and post-reform. The reforms were sufficiently exogenous, for they were not intended to directly sway the outcome of interest, which is the rates of non-communicable diseases that appeared years later. Further, the childhood years and the reforms were mutually independent so that neither was there a realistic way for the cohorts to self-select their decades of birth and the statistical bias such self-selection would entail; nor does it seem very likely that the laws were framed or timed to favor some cohorts of children and not others. A conditional finding from this experiment is that the era of the cohorts' childhood years—pre-reform or post-reform—may have mattered to the cohorts' profiles and thus to their aging. The parameters of the pre-reform and the post-reform cohorts turned out statistically distinct.

However, those distinctions stand apart from the estimates in Chap. 8, where the goal is to extract the childhood-linked signal in aging, using physiological growth, the state of women during pregnancy in the cohorts' birth-decades, inequality, per capita real incomes in the childhood years, and infections the cohorts had faced then. They reflect dimensions along which the cohorts' childhoods had differed other than being pre- or post-reform. Nonetheless, Chaps. 6 and 7 underscore a crucial point: while politics may not be as amenable to measurement as other variables, they can sway the direction of those variables, changing the outcome for aging years later.

That the politics may have influenced the path of modern aging and the associated healthcare spending, in turn, informs the discussion in Chap. 9 on the role of income elasticity. The elasticity estimated in the literature ranges widely: nearly zero or small at the micro level, about 1 at the regional level, and 1.5–1.6 at the aggregate-level. However, if resources allocated to healthcare have strong political roots, and institutions such as the National Health Service were set up so that incomes of individuals should not matter as much for accessing medical care, then estimates of income elasticity at the micro level would be irrelevant to the exercise in Eqs. 1.2, 1.3, and 1.4.

At the aggregate-level, too, if politics tell on healthcare allocations for a given path of real per capita income, income elasticity may not be construed in the usual sense of the percentage change in healthcare demand for a percentage change in real income. An example of this is the prospective costs of the NHS and the budgetary allocations estimated by Abel-Smith and Titmuss (1956), which, if had been adopted, would have set healthcare spending on a path independent of what any a priori income elasticity might have implied. In fact, it may have been responsible for some "income elasticity" but one that may not be construed in the usual sense. An even older example is that of the large public health infrastructure investments for containing infectious diseases toward the end of the nineteenth century and early twentieth century. A more recent example is the call by Simon Stevens, the chief executive of the NHS, for additional funding under the rubric of the Five Year Forward View. In this concurrent instance, as was in each of the prior cases, if resources are made available over a series of years, some income elasticity-like term would figure for the given future path of real per capita GDP, but it would not be interpretable as the usual income elasticity as defined at the micro level.

Further, if allocations in the future were to hinge on voter preferences and needs; on the political response to voter preferences; and on systemic economic dislocations like the most recent crisis in 2007–2008; then knowing the elasticity beforehand may not be useful at all for framing an outlook for healthcare spending. It could even be irrelevant for the purpose. Why? Both systemic dislocations and political changes are unforeseeable, and cannot be the baseline scenarios in any outlook decades out. For the purpose of policy, even if one assumes that an 'elasticity' is tangibly available, and, as a policy tool, can be constricted to rein in healthcare spending, it is likely to be a blunt and perilous way of containing costs: as the allocations shrink across the board and are relatively unresponsive to the changing underlying needs, one may not know where the damage might fall.

Overall, the evidence conveys that economic and political change are vital for grasping modern aging, and at the same time, evolution of healthcare spending may not be grasped well without knowing physiological and the epidemiologic aspects of aging since the nineteenth century.

Before proceeding, a few more things about the book. Chapter 4 addresses what looks like a paradox: if non-communicable diseases climb with age, should not the aggregate rate have grown as the population aged ever more in the twentieth century? Why has the aggregate rate fallen instead? It sounds like a paradox because its premise is a static upward-sloping profile. If several successive cohort-profiles—all upward-sloping—shift down, the period-profiles shift downward as well, and the aggregate rate associated with the period-profiles trends downward. Chapters 3 and 4 demonstrate that such shifts are key for understanding the trends of the aggregate rate, and thus of aging.

This issue has a parallel in the debate on healthcare spending. That healthcare expenses climb exponentially with age, all else equal, does not imply that spending will grow exponentially as the baby-boom generation ages in the years ahead (Reinhardt 2003). The notion that healthcare spending is set to soar has been popular since at least the 1980s. Even as recently as October 6, 2014, for instance, the *Financial Times* wrote in its editorial: "... since the NHS spends about seven times as much on 80-year-old person as on a 30-year-old and coming years will see rapid rise in the elderly share of the population, what politicians mean by real protection will feel like a 10% cut." In other words, left on its own, healthcare expenses will flare up as people age ever more because the expenses on 80-year-olds are multiples of the ones on 30-year-olds. The spending-profile is static. After illustrating the problem with this way of viewing the matter for non-communicable diseases in Chaps. 3 and 4, Chap. 9 revisits it in the context of healthcare spending, clarifying the point Fuchs (1990) raised about how the role of aging should be properly assessed.

Further, I am aware of the debate on living standards that engages some of the data the book explores (Steckel and Floud 1997). However, the book dwells only briefly on the topic, mostly in Chaps. 2 and 3, where it is clear that period-life expectancies can mislead inference about living standards. The mismatch is especially acute in the second half of the nineteenth century, when period-life expectancy was climbing while non-communicable diseases were growing even among the youth.

Moreover, there is a tradition in microeconomics that follows scholars such as Grossman (1972) to model health-related issues as health capital, akin to the conventional human capital models of schooling. The language used in the book—illness rates, aging and illness profiles, doubling times, compression, expansion, and so on—differs from the capital- investment- depreciation- and rate of return-like terms in the microeconomic tradition. All told, however, the language should not mask the spirit of the exploration, which is one of human abilities as envisioned in Schultz (1962) and of human capabilities as clarified by Sen (1984, 1999). How else may one view the lasting effects of children's development, or view the political inequality in the nineteenth century that left the masses less capable of giving their children a better start?

Finally, the findings in the book may not extend to other countries, and should not be generalized. There remain many unresolved matters that need more interdisciplinary communication and research.

References

Abel-Smith, B., & Titmuss, R. M. (1956). *The cost of the National Health Service in England and Wales*. London: Cambridge University Press.

Acemoglu, D., & Robinson, J. (2000). Why did the West extend the franchise? Democracy, inequality and growth in historical perspective. *The Quarterly Journal of Economics, 115*(4), 1167–1199.

Acemoglu, D., & Robinson, J. (2001). A theory of political transitions. *American Economic Review, 91*, 938–963.

Acemoglu, D., & Robinson, J. (2008). Persistence of power, elites and institutions. *American Economic Review, 98*(1), 267–293.

Allen, R. (2009). Engels' pause: Technical change, capital accumulation, and inequality in the British industrial revolution. *Explorations in Economic History, 46*(4), 418–435.

Atkinson, A. B., & Piketty, T. (2007). *Top incomes over the twentieth century: A contrast between continental-European and English-speaking countries*. Oxford: Oxford University Press.

Atkinson, A. B., & Piketty, T. (2010). *Top incomes: A global perspective*. Oxford: Oxford University Press.

Barker, D. J. P. (1992). *Fetal and infant origins of adult disease*. London: British Medical Journal.

Barker, D. J. P. (1994). *Mothers, babies and disease in later life*. London: British Medical Journal.

Barker, D. J. P. (1998). *Mothers, babies and health in later life*. Edinburgh: Churchill Livingstone.

Barker, D. J. P. (2004). The developmental origins of well-being. *Philosophical Transactions of the Royal Society: Biological Science, 359*(1449), 1359–1366.

Barker, D. J. P., & Lackland, D. T. (2003). Prenatal influences on stroke mortality in England and Wales. *Stroke, 34*(7), 1598–1602.

Barker, D. J. P., & Osmond, C. (1986a). Infant mortality, childhood nutrition and ischemic heart disease in England and Wales. *Lancet, 1*(8489), 1077–1081.

Barker, D. J. P., & Osmond, C. (1986b). Childhood respiratory infections and adult chronic bronchitis in England and Wales. *British Medical Journal, 293*(6557), 1271–1275.

Barker, D. J. P., Godfrey, K. M., Fall, C., Osmond, C., Winter, F. D., & Shaheen, S. O. (1991). Relation of birth weight and childhood respiratory infection to adult lung function and death from chronic obstructive airways disease. *British Medical Journal, 303*(6804), 671–675.

Baumol, W. J. (1967). Macroeconomics of unbalanced growth: The anatomy of urban crisis. *American Economic Review, 57*, 415–426.

Baumol, W. J. (1988, Fall). Containing medical costs: Why price controls won't work. *The Public Interest, 93*, 37–53.

Bengtsson, T., & Lundstrom, M. (2003). Airborne infectious diseases during infancy and mortality in later life in Southern Sweden, 1799–1894. *International Journal of Epidemiology, 32*(2), 286–294.

Ben-Shlomo, Y., & Davey Smith, G. (1991). Deprivation in infancy or in adult life: Which is more important for mortality risk? *Lancet, 337*(8740), 530–534.

Ben-Shlomo, Y., & Kuh, D. (2002). A life course approach to chronic disease epidemiology: Conceptual models, empirical challenges and disciplinary perspectives. *International Journal of Epidemiology, 31*(2), 285–293.

Bleakley, H. (2007). Disease and development: Evidence from hookworm eradication in the American South. *The Quarterly Journal of Economics, 122*(1), 73–117.

Breyer, F., Felder, S., & Costa-Font J. (2011). Does aging really affect health expenditures? If so, why? www.voxeu.org. Accessed 14 May 2011.

Costa, D. L. (2002). Changing chronic disease rates and the long-term declines in functional limitations among older men. *Demography, 39*(1), 119–137.

Costa, D. L. (2003). Understanding the mid-life and older-age mortality declines: Evidence from the Union Army Veterans. *Journal of Econometrics, 112*(1), 175–192.

Crimmins, E. M., & Finch, C. E. (2006). Infection, inflammation, height and longevity. *Proceedings of the National Academy of Sciences, 103*(2), 498–503.

Culyer, A. J. (1989). Cost containment in Europe. In *Healthcare systems in transition*. Paris: OECD.
Dasgupta, P. (1993). *An inquiry into well-being and destitution*. Oxford: Clarendon Press.
Davey Smith, G., Hart, C., Upton, M., et al. (2000). Height and risk of death among men women: Aetiological implications for association with cardiorespiratory disease and cancer mortality. *Journal of Epidemiology and Community Health, 54*, 97–103.
Davey Smith, G., Gunnel, D., & Ben-Shlomo, Y. (2001). Life-course approaches to socio-economic differentials in cause-specific adult mortality. In D. Leon & G. Walt (Eds.), *Poverty, inequality and health: An international perspective* (pp. 88–124). Oxford: Oxford University Press.
Deaton, A. (2013). *The great escape*. Princeton: Princeton University Press.
Elo, I. T., & Preston, S. H. (1992). Effects of early life conditions on adult mortality: A review. *Population Index, 58*(2), 186–212.
Financial Times. (2014, October 6). *Britain's health service needs honest treatment*, Editorial. www.ft.com
Finch, C. E., & Crimmins, E. M. (2004). Inflammatory exposure and the historical changes in human lifespans. *Science, 305*, 1736–1739.
Floud, R., Wachter, K., & Gregory, A. (1990). *Height, health, and history: Nutritional status in the United Kingdom, 1750–1980*. Cambridge: Cambridge University Press.
Floud, R., Fogel, R. W., Harris, B., & Hong, S. C. (2011). *The changing body, health nutrition and human development in the Western World since 1700* (NBER). New York: Cambridge University Press.
Fogel, R. W. (1993). Economic growth, population theory and physiology: The bearing of long-term processes on the making of economic policy. *American Economic Review, 84*(3), 369–395.
Fogel, R. W., & Costa, D. L. (1997). The theory of technophysio evolution with some implications for forecasting population, health care costs, and pension costs. *Demography, 34*(1), 49–66.
Fries, J. F. (1980). Aging, natural death and compression of morbidity. *New England Journal of Medicine, 303*(3), 130–135.
Fries, J. F. (1989). The compression of morbidity: Near or far? *Milbank Memorial Quarterly, 67*(2), 208–232.
Fries, J. F. (2005). The compression of morbidity. *Milbank Quarterly, 83*(4), 801–823.
Fuchs, V. R. (1990). The health sector's share of the Gross National Product. *Science, 247*, 534–538.
Gompertz, B. (1825). On the nature of the function expressive of the law of human mortality, and on the mode of determining the value of life contingencies. *Philosophical Transactions of the Royal Society, 115*, 513–585.
Gompertz, B. (1872). On the uniform law of mortality from birth to extreme old age, and on the law of sickness. *Journal of the Institute of Actuaries, 16*, 329–344.
Grossman, M. (1972). On the concept of health capital and the demand for health. *Journal of Political Economy, 80*(2), 223–255.
Harris, B. (2001). "The child is the father to man" the relationship between child health and adult mortality in the 19th and 20th centuries. *International Journal of Epidemiology, 30*, 688–696.
Harris, B., & Floud, R. (1997). Health, height, and welfare: Britain, 1700–1980. In R. H. Steckel & R. Floud (Eds.), *Health, welfare and industrialization*. Chicago: University of Chicago Press.
Johansson, S. (1991). The health transition: The cultural inflation of morbidity during the decline of mortality. *Health Transition Review, 1*, 39–65.
Jousilahti, P., Tuomilehto, J., Vartiainen, E., Eriksson, J., & Puska, P. (2000). Relation of adult height to cause-specific and total mortality: A prospective follow-up study of 31,199 middle-aged men and women in Finland. *American Journal of Epidemiology, 151*(11), 1112–1120.
Komlos, J. (1993a). Further thoughts on the nutritional status of the British population. *Economic History Review, 46*(3), 363–366.

Komlos, J. (1993b). A Malthusian episode revisited: The height of British and Irish servants in Colonial America. *Economic History Review, 46*(2), 768–782.

Komlos, J. (2008). On English pygmies and giants: The physical stature of English youth in the late 18th century and early 19th centuries. *Research in Economic History, 25*(2), 149–168.

Kuh, D., & Davey Smith, G. (2004). A life-course and adult chronic disease: An historical perspective with particular reference to coronary heart disease. In D. Kuh & Y. Ben Shlomo (Eds.), *A life-course approach to chronic disease epidemiology*. Oxford: Oxford University Press.

Kuh, D., Hardy, R., Lagenberg, C., Richards, M., & Wadsworth, M. E. J. (2002). Mortality in adults aged 26–54 years related to socioeconomic conditions in childhood: Post-war birth cohort study. *British Medical Journal, 325*(Nov.), 1076–1080.

Leon, D. A., Koupilova, I., Lithell, H. O., Berguland, L., et al. (1996). Failure to realize growth potential in utero and adult obesity in relation to blood pressure in 50-year old Swedish men. *British Medical Journal, 312*(7028), 401–406.

Leon, D. A., Lithell, H. O., Vagero, D., et al. (1998). Reduced fetal growth rate and increased risk of death from ischaemic heart disease: Cohort study of 15000 Swedish men and women born 1915–25. *British Medical Journal, 317*(7153), 241–245.

Lindert, P. (2000). Three centuries of inequality in Britain and America. In A. B. Atkinson & F. Bourguinon (Eds.), *Handbook of income distribution* (Vol. 1, pp. 176–316). Amsterdam: Elsevier Science.

Makeham, W. M. (1867). On the law of mortality. *Journal of the Institute of Actuaries, 13*, 325–358.

Newhouse, J. P. (1992, Summer). Medical care cost: How much welfare lost? *Journal of Economic Perspectives, 6*(3), 3–21.

OECD. (2013). *Public spending on health and long-term care: A new set of projections* (OECD economic policy papers, no. 6). Paris: OECD.

Oliveira Martins, J., & de la Maisonneuve, C. (2006). The drivers of public health expenditure on health and long-term care: An integrated approach. *OECD Economic Studies* No. 43, 2006/2.

Olshansky, J., & Ault, A. B. (1986). The fourth stage of the epidemiologic transition: The age of delayed degenerative diseases. *Milbank Memorial Fund Quarterly, 64*(3), 355–391.

Olshansky, J., & Carnes, B. A. (1997). Ever since Gompertz. *Demography, 34*(1), 1–15.

Omran, A. R. (1971). The epidemiologic transition: A theory of epidemiology of population change. *The Milbank Quarterly, 49*(4), 509–538.

Omran, A. R. (1982). Epidemiologic transition. In J. A. Ross (Ed.), *International encyclopedia of population* (pp. 172–183). London: The Free Press.

Osmani, S., & Sen, A. K. (2003). The hidden penalties of gender inequality: Fetal origins of ill-health. *Economics and Human Biology, 1*(1), 105–121.

Paneth, N., & Susser, M. (1995). Early origins of coronary heart disease (the Barker Hypothesis). *British Journal of Medicine, 310*(6977), 411–412.

Piketty, T. (2014). *Capital in the twenty-first century*. Cambridge, MA: The Belknap Press of Harvard University Press.

Reinhardt, U. E. (2003). Does the aging of the population really drive the demand for health care? *Health Affairs, 22*(6), 27–39.

Schultz, T. W. (1962). *Investment in human beings*. Chicago: University of Chicago Press.

Schwartz, W. B. (1987, January). The inevitable failure of cost containment strategies: Why they can provide only temporary relief. *Journal of American Medical Association, 257*, 220–224

Sen, A. K. (1984). *The living standard*. Oxford Economic Papers, 36, U.K.

Sen, A. K. (1999). *Commodities and capabilities*. Oxford: Oxford University Press.

Solow, R. M. (1956). A contribution to the theory of economic growth. *The Quarterly Journal of Economics, 70*(1), 65–94.

Steckel, R. H. (1979). Slave height profiles from coastwise manifests. *Explorations in Economic History, 16*(4), 363–380.

Steckel, R. H. (1995). Stature and the standard of living. *Journal of Economic Literature, 33*(4), 1903–1940.

Steckel, R. H., & Costa, D. L. (1997). Long-term trends in health, welfare and economic growth in the United States. In R. H. Steckel & R. Floud (Eds.), *Health, welfare and industrialization*. Chicago: University of Chicago Press.

Steckel, R. H., & Floud, R. (1997). *Health, welfare and industrialization*. Chicago: University of Chicago Press.

Szreter, S. (1997). Economic growth, disruption, deprivation, disease, and death: On the importance of politics of public health for development. *Population and Development Review, 23*(4), 693–728.

Szreter, S. (2005). *Health and wealth: Studies in history and policy*. Rochester: University of Rochester Press.

Troesken, W. (2004). *Water, race and disease*. Cambridge: MIT Press.

United Nations Organization. (2012). Political declaration of the high-level meeting of the general assembly on the prevention and control of non-communicable diseases. www.un.org, document A/66/L.1.

Vaupel, J. W. (2010). Biodemography of aging. *Nature, 464*, 536–542.

Waaler, H. T. (1984). Height, weight and mortality: The Norwegian experience. *Acta Medica Scandinavia, Supplement, 679*, 1–51.

Weisbrod, B. A. (1991, June). The healthcare quadrilemma: An essay on technological change, insurance, quality of care, and cost containment. *Journal of Economic Literature, 29*, 523–552.

Werblow, A., Felder, S., & Zweifel, P. (2007). Population ageing and health care expenditure: A school of 'red herrings'? *Health Economics, 16*, 1109–1126.

World Health Organization. (2014). *Non-communicable diseases country profiles*. www.who.int. Geneva: WHO Press.

Zweifel, P., Felder, S., & Meiers, M. (1999). Ageing of population and health care expenditure: A red herring? *Health Economics, 8*, 485–496.

Chapter 2
The Epidemiologic Transition

2.1 Seen as Regimes

The Transition has traditionally been portrayed as a series of regimes that have distinct features. Exploring the regimes in the data inevitably means that various periods will have to be labeled, and a rough judgment has to be made about which period is more akin to which regime. The goal of this chapter is to explore the main features of the regimes, and highlight the epidemiologic situation cohorts of children brought up in them had received. The macroeconomic and political backdrops accompanying the epidemiologic one discussed in this chapter are outlined in Chap. 6. Further, the features of the regimes have traditionally been outlined as 'propositions,' which are like stylized facts that help keep track of the Transition's main developments. The chapter follows the tradition and states as propositions some of the facts vital for grasping the trends of non-communicable diseases over the very long term; they form the basis for the exploration of aging later in the book.

The three regimes outlined in Omran (1971) are the Malthusian-type 'age of pestilence and famine;' the transition 'age of receding pandemics;' and the post-transition 'age of degenerative and man-made diseases.' It is important to keep in mind that the 'Malthusian' label refers to the epidemiologic situation, not necessarily the economic one.

In the Malthusian-type regime, the main causes of deaths are infectious diseases. Their epidemics recur frequently, often developing into pandemics. Period-life expectancies are low and volatile, indicating that average lives are likely to have been shorter and more uncertain than in other regimes. Especially vulnerable are period-life expectancies at early ages as the regime is harsh on children. Recurring famines, pestilences, and natural disasters foster hunger and malnutrition, beset children's physiological growth, often leaving them stunted.

Then, for various reasons, infectious diseases start falling in the transition 'age of receding pandemics.' While the masses benefit from this new development,

S. Arora, *The Transitions of Aging*, International Perspectives on Aging 12,
DOI 10.1007/978-3-319-14403-0_2

children gain more than others. People on average are able to live longer. Consequently, period-life expectancies grow, more at younger ages than at older ones.

As infectious diseases recede and lives extend, people become increasingly prone to non-communicable diseases that surface mostly in the adult stages of life. Such diseases gradually become the main causes of sickness and deaths in the post-transition 'age of degenerative and man-made diseases.'

The main tenet of this three-stage framework, stated as Proposition 2 in Omran (1971), says:

> During the transition, a long-term shift occurs in mortality and disease patterns whereby pandemics of infection are gradually displaced by degenerative and man-made diseases as the chief form of morbidity and primary cause of death.

In the 1970s, when the framework was proposed, infectious diseases had long been subdued; but the rates of many non-communicable diseases, especially in the middle-to-older age people, had stayed high. The situation was expected to persist, for an increasingly aging population could only keep the rates of such diseases elevated, and keep period-life expectancies from growing much further. The evidence until then had signaled such a prospect as well: over the first half of the twentieth century, life expectancy for, say, 65–74-year olds had inched up from 10.3 years to just 11.9 years, as if it had run into some stubborn limit (HMD).

However, new trends in the 1960s and 1970s took scholars by surprise: instead of rising, many non-communicable diseases among the middle-to-older age people began falling. Consequently, period-life expectancies began surpassing the erstwhile limits that had been proposed (Oeppen and Vaupel 2002).

In the wake of those new trends, a fourth stage called the 'age of delayed degenerative disease' was added to the original three-stage framework (Omran 1982; Olshansky and Ault 1986). The issues central to this new ongoing stage, and to the debate on aging, have been: (1) non-communicable diseases that accompany aging; (2) the prospects of such diseases as people age more in the upcoming decades; and (3) the consequences for period-life expectancies, especially at middle-to-older ages, if those prospects become actuality.

The scenarios of aging outlined in the previous chapter have been developed from this debate on the prospects of non-communicable diseases. On the one side of the debate, scholars have argued that the falling rates of such diseases should be seen as transient or as mere delays. More aging makes people only more prone to such diseases, and the rates would eventually climb (Olshansky and Ault 1986; Carnes and Olshansky 1993). Checked by the finite limit to lifespan, then, period-life expectancies would run into some new limit as well. Running into the limit also means that ongoing delays would lead to the compression of deaths (Fries 1980, 1989, 2005). Meanwhile, the other side of the debate has argued that there does not appear to be an a priori limit to period-life expectancies, implying that rates of non-communicable diseases may continue to shrink instead (Oeppen and Vaupel 2002). After exploring the evidence on the age-profiles of non-communicable diseases, and the likely trends the profiles imply, later chapters will return repeatedly to this central issue of the debate.

Meanwhile, the proposition cited above speaks of long-term 'shifts.' This chapter clarifies whether they pertain to the rates (levels) of the two disease categories, or to their shares of the total. For instance, the shares can stay stable as the rates fall; or, the share of one category can expand while that of the other can shrink, even as the rates of both categories fall. Isolating the levels from the shares would clarify the matter.

2.2 And in Data Over at Least Two Centuries

To explore the regimes properly, data over a very long period are required. The data would ideally run for a long-enough spell of the Malthusian regime and then cover the transition from that regime to the modern one. The ideal, however, is beyond reach as comprehensive data on various disease categories has been amassed only since Registration became well established circa mid-nineteenth century. The data since then would cover barely two to three decades of what appears to have been a Malthusian-type regime; so other types of evidence will need to be brought in to fill in the gap as much as possible.

Further, data on disease-caused deaths since circa 1850 are useful only after they have been formed into consistent series by following the International Classification of Diseases (ICD). The ICD lists specify diseases that make up the two broad categories mentioned in the proposition. In the early decades after the onset of Registration, some of the basic principles of the classification had been set and adopted officially. Most of those principles have since been the bedrock of the ICDs issued first by the World Statistical Congress, then by the League of Nations, and finally by the World Health Organization. The continuity offered by the ICDs helps make the series as consistent as possible over time, though the revisions to the lists make the task tedious and the outcome less than perfect.

To rebuild the time series of various disease categories, scholars usually adopt a recent ICD list (e.g., ONS 1997). Following that lead, the book adopts the ninth revision (ICD-9, WHO 1978) and isolates since the mid-nineteenth century the first main category: all infectious diseases including diseases of the respiratory system, which is mostly made up of infections such as influenza, pneumonia and bronchitis.

After isolating most infectious diseases, the series for the non-communicable category is calculated as a residual aggregate thus: all causes minus all external causes such as accidents, homicides, injuries etc., minus complications of pregnancy and childbirth, minus all causes covered under the infectious disease category mentioned above. This aggregate is made up of non-communicable sub-categories such as the diseases of the circulatory system (e.g. heart), the nervous system (e.g. brain), the digestive system (e.g. liver), the musculoskeletal system (e.g. bones), and neoplasms (various cancers). The book will mainly study aggregate non-communicable diseases, but will at times examine some of its sub-categories. The appendix lists the compositions of the sub-categories, outlining the corrections to their reported totals necessary for making them as consistent as possible. It is important to note the caveats mentioned in the appendix. The most important one is that perfect alignment of particular diseases may be nearly impossible across the ICDs. Therefore, it is safer to study broad groups. Broader the groups of diseases considered, greater the chances of cross-time consistency. In that sense, one can be more confident about the consistency of the broadest groups—infectious diseases, non-communicable diseases—than about their narrower sub-categories.

2.3 That Show the Main Features of the Regimes

Figure 2.1a shows that period-life expectancy at birth had fluctuated between 27 and 40 years for at least three centuries. It permanently surpassed the range only in the last few decades of the nineteenth century, a change of direction also visible in the less-inclusive period-life expectancies at ages 15 and 20 in Fig. 2.1b. Together they suggest that stimuli unique to the late nineteenth century had likely ushered-in a new era that lasted through circa 1940. Roughly 70–75% of the total gain over 1871–2000 had already been achieved by then.

The main proximate reason for the gain was the decline of infectious diseases. Marked by significant outbreaks, their path in Fig. 2.2 shows a permanent change of direction since about the 1870s. Not all such diseases began receding then, but a large segment, especially respiratory infections, had changed direction in the last two to three decades of the nineteenth century.

Figures 2.1 and 2.2 suggest roughly the periods that could be considered as regimes. The period before circa 1870 has Malthusian era-type traits (outlined below); 1870–1920/1930 the transition stage; after then, the post-transition one.

Frequent famines, one of the main features in the received view of the Malthusian regime, had largely been subdued by the start of the nineteenth century. It was neither less variability of weather that had changed the situation; nor were there fewer natural disasters than before. Famines were subdued by better policy of food relief that "had unalterably shifted back to the ideas and practices of common wealth that had prevailed during 1600–1640" (Fogel 1993; Floud et al. 2011, pp. 116–118).

However, subduing famines only implies that acute crises of hunger, triggered usually by soaring food prices and entrenched entitlements, may have been checked by better policy than before (Sen 1981). It does not necessarily mean that enough food was available for all during normal times. Chronic hunger could still have persisted. For instance, about 20% of the British population at the start of the nineteenth century had access to fewer than 2,500 calories per consuming unit (within households) per day, barely enough to support the basic metabolic rate and some light work (Floud et al. 2011, pp. 167–169). It is now established that real wages grew very slowly over 1770–1830, and even paused over a significant segment of the period (Allen 2009). Stagnant real incomes and insufficient

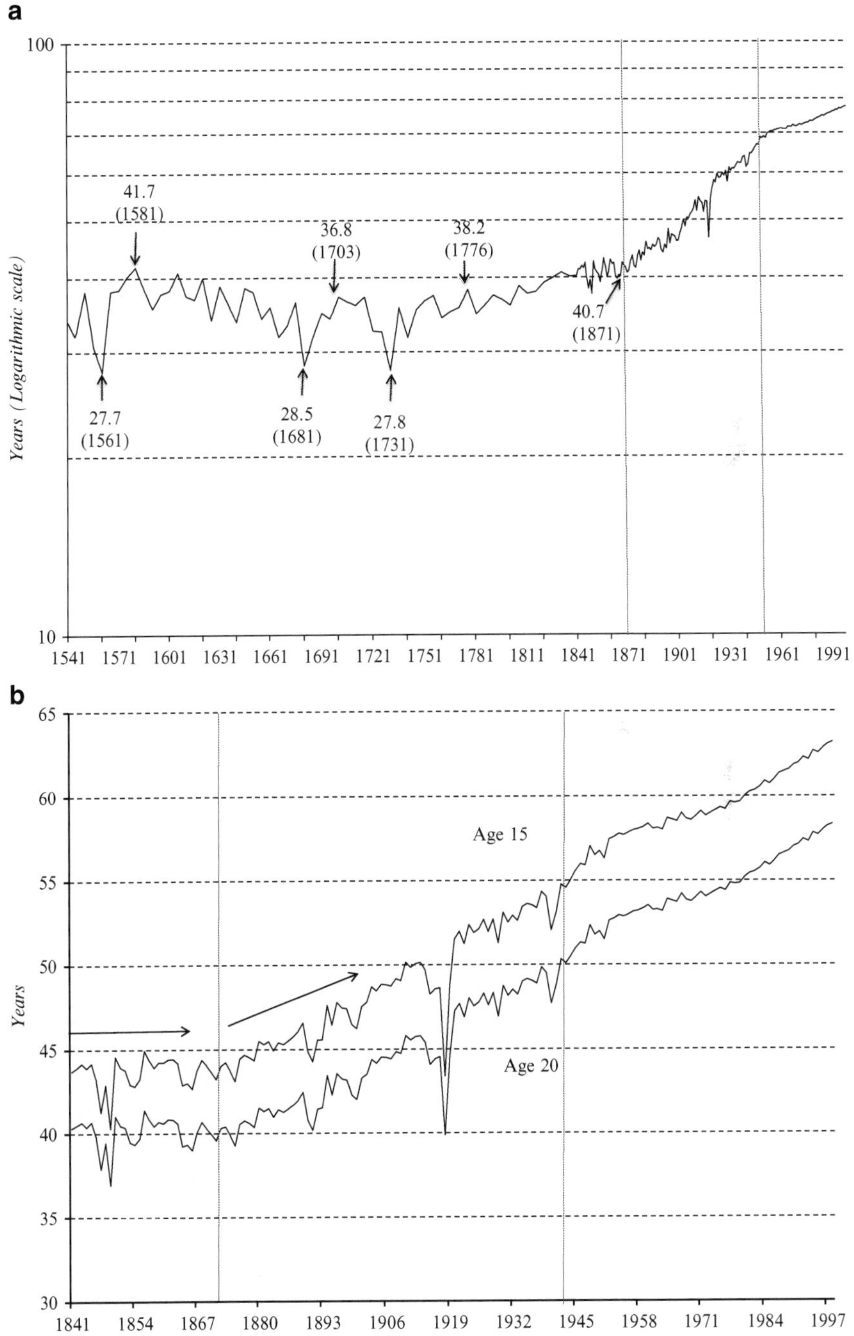

Fig. 2.1 Period-life expectancy (**a**) At birth 1541–2000. (**b**) At age 15 and 20, England and Wales 1841–2000 (Sources: Wrigley and Schofield (1981), Human Mortality Database)

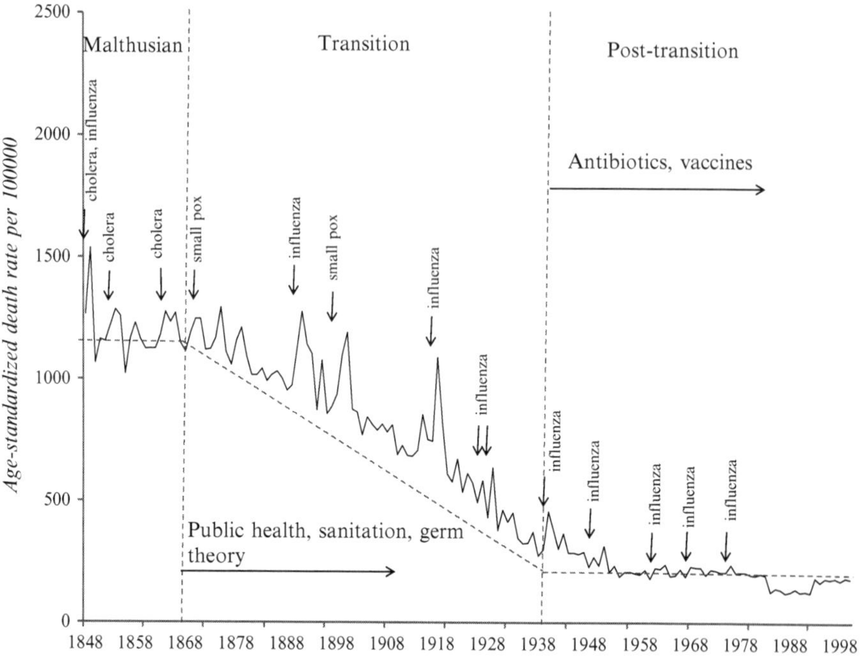

Fig. 2.2 Infectious diseases, 1848–2000 (Sources: HMSO; author's calculations)

food may have also forced families to reallocate more calories to the breadwinners, and that could have resulted in the poor nutritional status of working-class women and children (Harris 1994, 2004b; Horrel and Oxley 1999; Oxley and Meredith 2014), who, in the event the male breadwinner died, often had to fend for themselves at even lower wages: "..when male wages were not forthcoming or insufficient for family support (and these occasions were many and varied) children were cast as secondary earners in the typical working-class family" (Humphries 2013, p. 85).

While chronic hunger, not acute crises such as famines, may have been the more relevant feature during the first half of the nineteenth century, the epidemiologic side of the picture is much closer to the Malthusian-type portrayed in the received view. Figure 2.1 shows that period-life expectancy at birth, the most inclusive of life expectancies, had persisted between 27 and 40 years before the 1870s. Though it had trended up mildly since the mid-eighteenth century, its path had stayed quite volatile.

Over 1541–1870, it averaged 36.4 years with a standard deviation of 3.1 years. However, there seem to be two distinct segments in those 330 years. The average over the first segment, 1541–1740, was 35.4 with the standard deviation of 3.1 years. The new average, coinciding roughly with the Industrial Revolution 1741–1870, was 38 years with a standard deviation of 2.5 years. This new average was within one standard deviation of the old one, which means that the new outcomes of period-life expectancy would be expected within two-thirds of all outcomes in the prior two centuries as well. However, the old average was slightly less than one standard deviation of the new, or would be expected in only about a third of the outcomes over 1741–1870. On that basis one may infer a slight step-up from the past, though still not decisively.

Life expectancies over the first half of the nineteenth century were grimmer in urban areas than at the national level. Over 1820–1870, the national-level life expectancy at birth was 40–41 years. The ones in major provincial cities were 35 years in the 1820s, nearly 29 in the 1830s, 30 in the 1840s, and about 34 in the 1850s and 1860s (Szreter 2005). In the 1850s, for instance, the registration districts of the largest cities, defined as ones with more than 100,000 people by 1851, had expectancies ranging from 40 to 42 in Clifton and Aston, about 34 in Sheffield and Leeds, to as low as 27 in Liverpool. In the 1860s, the best of them posted 42 years; the worst, in Liverpool, had shrunk to 25—a level last seen during the Black Death (Szreter and Mooney 1998; Szreter 2005).

Further, Fig. 2.1b shows that the national-level life expectancies at ages 15 and 20 had not changed much over 1841–1870. At age 15, it was 43.7 years in 1841 and 43.7 years in 1870. For 20-year-olds, it was 40.3 years in 1841 and 39.6 years in 1870. All told, life expectancies may have advanced somewhat, but not convincingly.

The next feature of this regime is the rates of infectious diseases and their outbreaks. Comprehensive data on disease-caused deaths before circa 1850 are unavailable. Figure 2.3 may offer a sense of the epidemiologic situation in the first half versus the second half of the nineteenth century. It shows how frequent epidemics were over the 50-year periods since the seventeenth century. To make the bar graphs, each recorded outbreak in a country was assigned a value of 1. If the outbreak was a pandemic, then each country involved was allotted 1. The tally for each country was then summed over 50-year periods to yield a within-country view.

The greater frequency of epidemics in the first half of the nineteenth century than in other periods indicates that the epidemiologic situation may have been worsening as urban areas were becoming increasingly densely populated during the Industrial Revolution (Wohl 1983, p. 4; Szreter 1997, p. 64; Woods 1985, pp. 360–380). Some of the outbreaks in Britain alone include influenza 1803, smallpox 1816–1819, typhus 1816–1819, dysentery 1817–1819, smallpox 1825–1826, influenza 1830, typhus 1837–1838, smallpox 1837–1840, typhus 1847–1848, dysentery 1847–1850, influenza 1847–1848, and the infamous cholera outbreaks in 1848–1849, 1853–1854, and 1865–1866.

That type of pattern is visible in the other countries shown in Fig. 2.3. Historians of public health have called the first three-quarters of the nineteenth century as the era of pandemics. For instance, cholera epidemics appeared in Asia, Europe and North America over 1826–1837; influenza 1836–1837 had spread in Asia, in Europe, and in North America; malaria in 1823–1827, and influenza in 1847–1848, had spread through Europe; over 1853–1859, cholera surfaced in Britain, France, Germany, Canada, and Spain; the 1865–1866 cholera had struck Britain and France, among other countries in Europe. Expressing concern about the recent

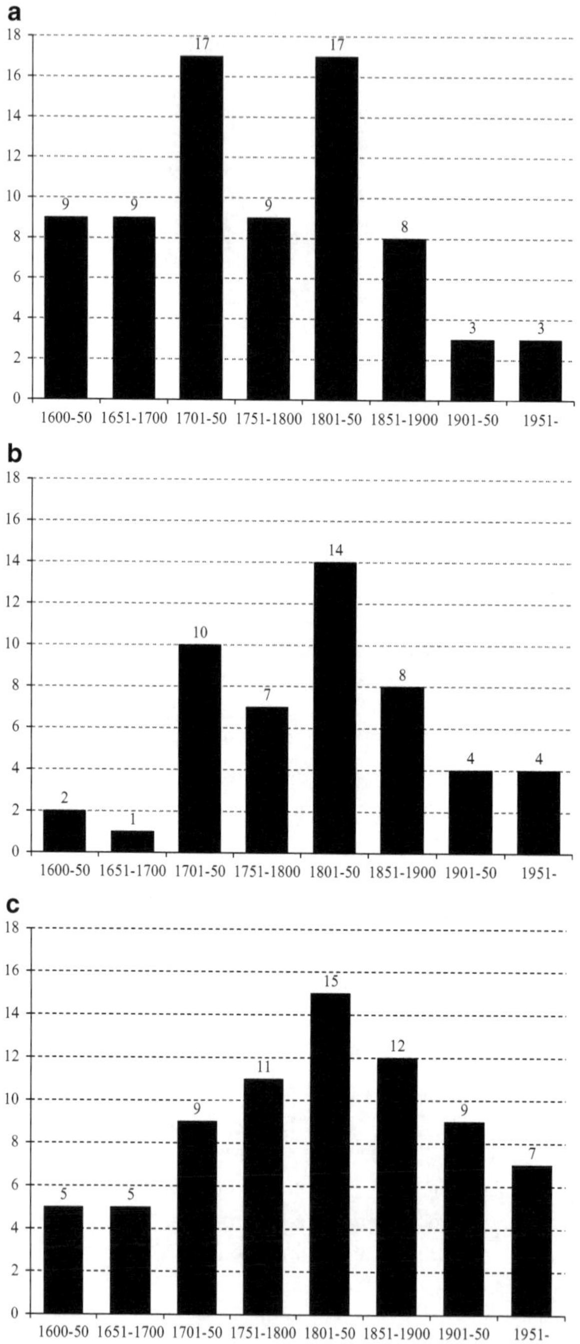

Fig. 2.3 The frequency of epidemic outbreaks, 1600–2000 (**a**) Britain & Ireland (**b**) France (**c**) USA/Canada (**d**) Italy/Spain (**e**) Germany/Prussia/Austria (Sources: Kohn (1998); author's calculations)

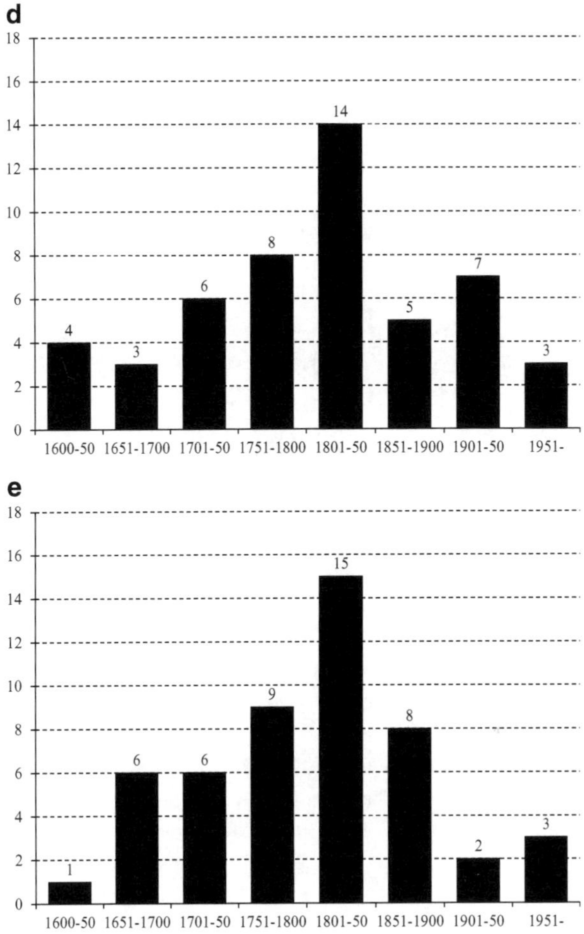

Fig. 2.3 (continued)

Ebola outbreak, and about new emerging infections, Jeffrey Sachs (2014), for instance, correctly states, "These epidemic diseases are new markers of globalization, revealing through their chain of death how vulnerable the world has become from the pervasive movement of people and goods." However, as Fig. 2.3, and the list of cross-continental pandemics in the nineteenth century reveal, infectious diseases have been one of the oldest markers of the movement of people and goods. It took institutional reform (across countries) and new knowledge to contain such diseases while letting globalization proceed.

Though the pattern in Fig. 2.3 indicates that the rates of infectious diseases are likelier to have been more in the first half of the nineteenth century than in the second, the inference would still be indefinite without comprehensive data. Only for the two to three decades for which there is such data may one say with some

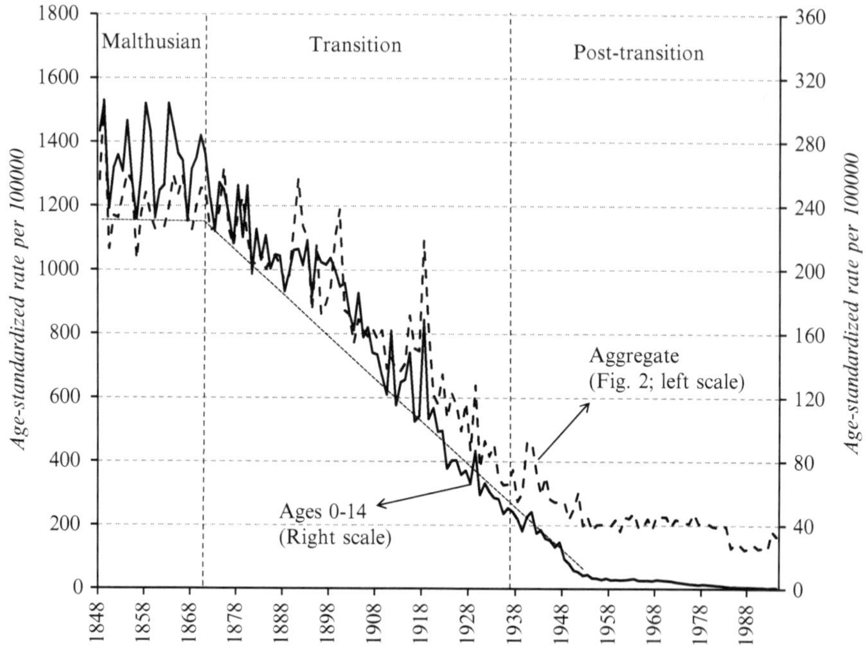

Fig. 2.4 Long-term trends of infectious diseases in aggregate and in children aged 0–14, 1848–2000 (Sources: HMSO; author's calculations)

confidence that poor period-life expectancies had coincided with very high rates of infectious disease (Fig. 2.2): the rates in the 1850s and the 1860s were some of the highest on record; about three to four times more than the rates in the 1930s, the cusp of the antibiotic revolution.

Figure 2.4 shows that the long-term trends of infectious diseases in children (defined as age 0–14 years) had followed a path similar to the one of population-wide rates. This long-term path likely traces the epidemiologic situation successive cohorts of children had received over the transition. To differentiate the cohorts born in the various decades, Chaps. 7 and 8 will use the rates of infectious diseases in children along this path as one of the variables reflecting childhood development across cohorts.

Moreover, some of the details along this path convey that children who grew up in the Malthusian stage received an extraordinarily harsh epidemiologic situation. For each decade on the horizontal axis of Fig. 2.5, the bars show children's share in aggregate infectious diseases. The first bar labeled 'raw' is children's shares unadjusted for their weight in the population. Consider the years 1855 and 1865 when children were about 35% of the population, and the aggregate population-wide rates were three to four times the rate seen in the 1930s and 1940s. In 1855, children accounted for 72.3% of diarrhea and dysentery, 48.1% of cholera, 39% of gastritis, enteritis and jaundice, 48% of influenza, pneumonia and bronchitis, 93.2% of whooping cough and scarlet fever, 21.5% of typhus, 68.8% of smallpox, 21.7%

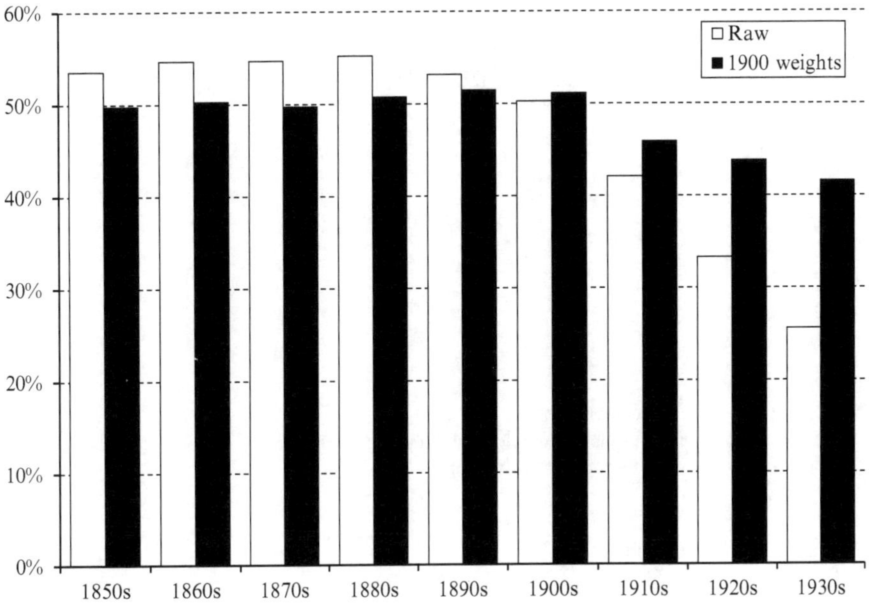

Fig. 2.5 Infectious diseases in children as a share of infectious diseases population-wide (Sources: HMSO; author's calculations)

of scrofula, and 61% of cephalitis. In 1865, they had 97% of all deaths from scarlet fever and whooping cough, 49% of influenza, bronchitis and pneumonia, 82.8% of diarrhea and dysentery, about 49% of gastritis, enteritis and jaundice, 20% of typhus, 65% of smallpox, 17% of scrofula, 57% of cholera and about 70% of cephalitis. Not only were the absolute level, or the rates, very high, children accounted for a disproportionately high share as well.

For some of the diseases, the high percentages in children would be expected because children usually do account for the larger share of diseases such as measles, chickenpox, whooping cough, scarlet fever, among others. Notable about the regime, however, is the unduly large share in diseases such as pneumonia, bronchitis, influenza, cephalitis, cholera, enteritis, among others that affect adults as well. Modern tallies do not show such high percentages, if deaths from such diseases occur at all. In 1948—to pick a year—children accounted for 29% of dysentery, 12% of influenza, 3.5% of bronchitis, and 30% of pneumonia; none from cholera, jaundice, enteritis, diarrhea, among a long list of others that seem to have been subdued by then.

For cross-decade comparisons, now consider the set of bars labeled '1900 weights' in Fig. 2.5. They are based on an age-adjusted rate calculated by using the age-distribution in 1900, when children's weight in the population was about 33%. Children had account for roughly 50% of all infectious diseases through the 1850s and the 1860s. Even as annual *rates* had been falling over the transition (Fig. 2.4), children continued to account for a sizable share, making up less than 30% only since about the 1940s.

Table 2.1 The heft of infectious diseases in children, age 0–14

Decade	Infectious disease rates	Ratio of infectious diseases to all causes	Ratio of infectious diseases to non-communicable diseases
1850s	266.6	0.6	1.6
1860s	263.2	0.6	1.6
1870s	245.2	0.6	1.7
1880s	210.6	0.6	1.6
1890s	205.5	0.6	1.6
1900s	165.1	0.6	1.5
1910s	129.7	0.6	1.6
1920s	84.7	0.6	1.3
1930s	52.7	0.5	1.0
1940s	31.8	0.4	0.7

Sources: HMSO; author's calculations
Note: The rates in the first column are calculated by averaging the age-standardized rates in each decade (using 1900 weights for standardization)

Table 2.1 shows that infectious diseases had likely been pertinent to children's epidemiologic situation in all stages even though the extent of the problem may have differed across the stages. The average rates echo the long-term trends shown in Fig. 2.4. Though the rates were falling, the next column shows that infectious diseases had accounted for roughly 60% of all causes in children through the turn of the century, and that share fell to less than 50% only in the 1940s. The last column shows that the rates of infectious diseases in children were 1.6–1.7 times the rates of various non-communicable ones, indicating that infections were the more significant category of diseases afflicting children. All told, Figs. 2.4 and 2.5 and Table 2.1 together indicate that infectious diseases should be one of the vital elements of the epidemiologic situation children had received during the Transition.

An insight into children's state comes from the average heights they were able to attain. Height attained by adulthood summarizes the net nutritional status, a net outcome of diets that fuel growth from infancy through adolescence, and the demands by infections that hinder physiological growth (Tanner 1990; Floud et al. 2011). Figure 2.6 shows the average height for children who had endured to age 18, by which time 99–100% of the final height had been attained (Floud et al. 1990, p. 188). Heights began growing mildly in the late eighteenth and early nineteenth century, but reversed for a few decades thereafter. They fell to as low 5 ft 3 in. for the cohorts born during the 1840s and 1850s. Chapter 6 will delve deeper into this trend by outlining the large gaps in the average heights of children across the social classes.

Meanwhile, at the nadir of the long-term trend, adult heights of the working-class children from the 1840s and 1850s (about 5 ft 3 in.) were less than the first centile of the modern height distribution. It means that an average working-class adult entering the work force stood only as tall as early teenage boys from the 1990s. Average heights have since improved, though it was not until the cohorts

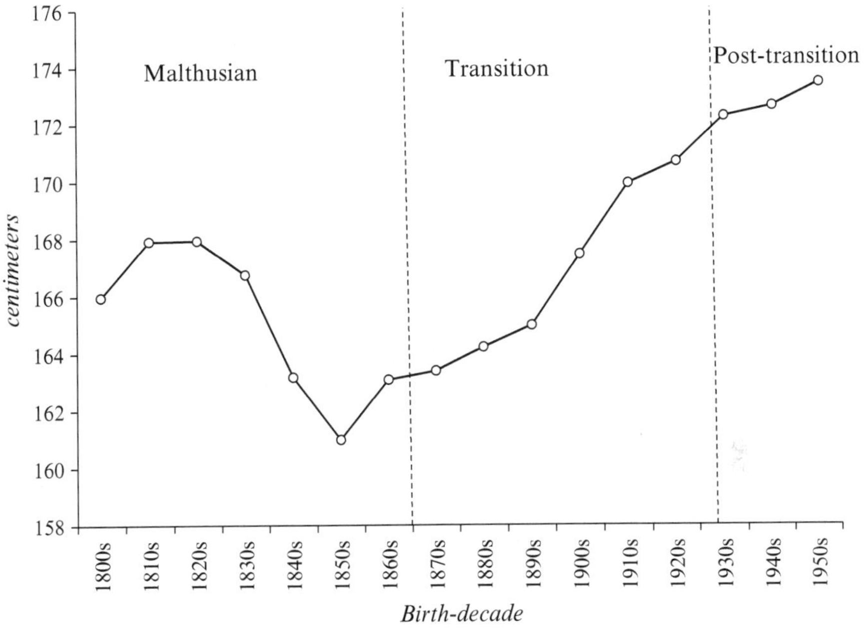

Fig. 2.6 Average adult heights of cohorts of children, 1790–1950 (Sources: Floud et al. 1990, Rosenbaum 1988)

born in the late nineteenth century that the prior downturn in heights of cohorts had been fully reversed. The average in the 1870s stood at roughly 5 ft 4 in., eventually reaching about 5 ft 8 in. for the 1940s cohort (Rosenbaum 1988; Floud et al. 1990). Some of the biggest gains toward the making of the modern height distribution were achieved during the transition stage.

To recap the situation before circa 1870, life expectancy at birth was low, volatile, and struggled to surpass levels seen in prior centuries; epidemic outbreaks were more frequent than in other regimes; life expectancies in urban areas were less than the ones at the national level; the ones after birth such as age 15 and 20 stayed stubbornly low; the aggregate rates of infectious diseases were the highest on record; children accounted for a hefty segment of it; and average adult heights successive cohorts were able to attain were near the bottom of the modern height distribution.

The lack of proper knowledge may have been one of the major reasons for those adverse developments. For instance, while the vivid symptoms of infectious diseases may have long been known, what caused such diseases to spread may not have been well understood. At least through the 1860s, the miasma theory appears to have been the main paradigm. One clue about its standing in official affairs appears in the Registrar General's Annual Statistical Reviews that had adopted the classifications of diseases issued by the Royal College of Physicians of London. Unlike the modern lists, the Reviews had classified diseases such as influenza, dysentery, typhoid, malaria, cholera, diphtheria, diarrhea, typhus, scarlet fever, among others, as miasmatic diseases. Tuberculosis was a 'constitutional' disease, not a contagious one as we know it today. Jaundice and hepatitis were 'local,' not infectious (see early tables by HMSO; Farr 1885). A growing consensus among experts at the time had been striving for public sanitation even before the Public Health Act of 1848 (Flinn 1965 pp. 18–43; Harris 2004a, p. 110). However, their efforts had faced opposition for at least two decades, held back by politics and by inadequate allocation of resources to the issue (Szreter 1988). Chapter 6 will discuss some of the main political changes, highlighting key events that may have helped usher-in the transition stage.

Meanwhile, the period that seems to fit best the description of the transition stage or the 'age of receding pandemics,' is roughly 1860/1870 through 1920/1930. Life expectancies made lasting gains during this time. The one at birth grew faster than the one at, say, age 20. Starting at about 41 years in 1871, expectancy at birth grew at an annual compounded rate of 0.6%, adding 20.2 years through 1940. It then expanded at 0.4% annual clip to reach 77.6 in 1998. The one at age 20 stood at 39.6 years in 1871, grew at 0.3% annual rate, adding 8.1 years through 1940, and then expanded 0.35% per year to attain 58.3 years in 1998.

Further, infectious diseases fell permanently since about the 1870s and have stayed subdued after circa 1940s. Their reduction over the transition stage accounts for about 95% of the total decline from the 1870s through 2000. Most large outbreaks receded as well, leaving only the air-borne influenza. The progress occurred even as urban areas became more densely populated than in the Malthusian regime. Multiple sources likely contributed to their fall though not all of them may have occurred at the same time. They include new arterial systems of water supply and sewage disposal; pasteurization, sterilization and antiseptic procedures; dissemination of preventive information as the new paradigm, the germ theory of diseases, became more widely known; and the consumption of more calories as incomes grew. Influential among the social choices were a series of Public Health Acts in 1866, 1872, and in 1875 that fortified the administrative set up for the long-term change (Flinn 1965; Frazer 1950; Woods 1985; Szreter 1988, 1997, 2005; Easterlin 1999). The role of sanitary reforms grew further at the turn of the century when the scale of the public health investments rose significantly (Harris 2004a p. 405; Floud et al. 2011, pp. 173–74; Chap. 9, this book).

To recap the situation since the 1860s/1870s, life expectancies at younger ages, especially at birth, made lasting gains, adding more years in the transition stage than in any other period; infectious diseases retreated and much of the progress against them was made in the decades since the 1870s; large epidemic outbreaks became far less frequent, as did pandemics; infectious disease rates in children fell, tracking broadly the long-term direction of the aggregate rate; children's average heights grew sizably as their physiological development improved.

2.4 As well as of the Paradox of Aging

Figure 2.7 shows a facet that the received view of the Transition may not have covered adequately: the sizable share of non-communicable diseases in the Malthusian regime. Circa 1850, they were about 50% of the raw total of all disease-caused deaths; about 55% based on 1994 standard weights; roughly 45% based on 1900 weights; and about 50%, based on the geometric mean of the last two. Further, one segment of this category, ill-defined causes, was about 23% of all disease-caused deaths through the early 1870s. Following ICD-9, the bulk of it was 'old age and senility,' a degenerative condition in people aged 65 years or older.

While the relative shares of the two broad disease categories would differ if the age-distribution from another year were used to calculate them, the data broadly convey that the so-called diseases of aging were large contributors in the Malthusian regime as well. Hence, the following proposition:

Proposition 1 *In the Malthusian regime, both infectious diseases and non-communicable diseases contributed sizably to all disease-caused deaths.*

In Omran's Proposition 2, non-communicable diseases had 'displaced' the infectious ones. Figure 2.7 clarifies that it was the *share* of the non-communicable diseases that 'displaced' the share of the infectious ones.

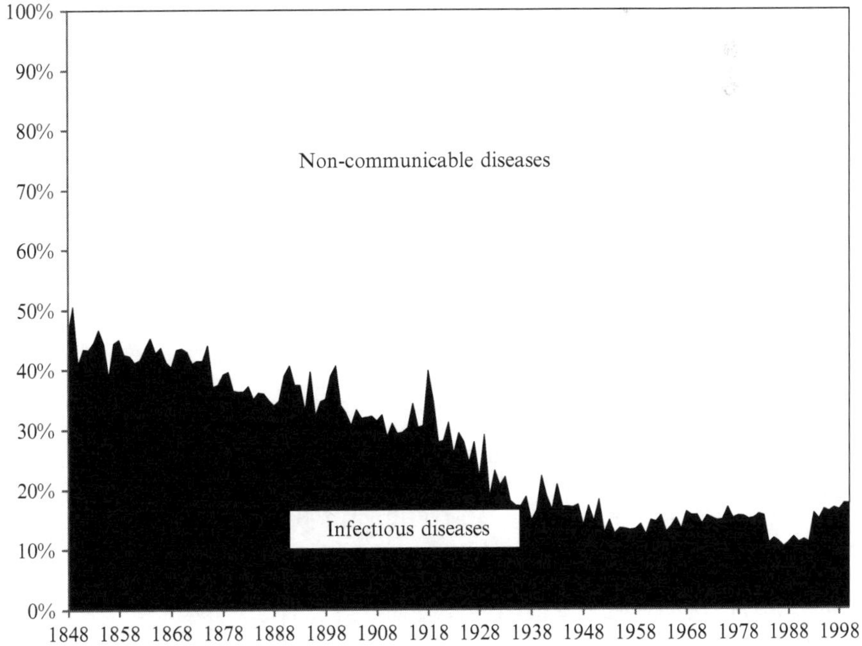

Fig. 2.7 Infectious diseases and non-communicable diseases as percentages of all diseases, 1848–2000 (Sources: HMSO; author's calculations)

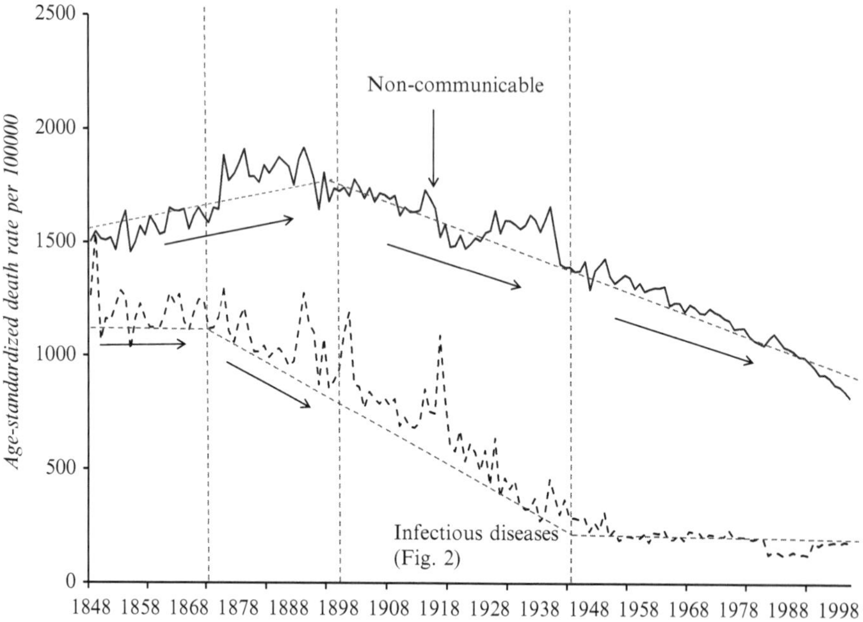

Fig. 2.8 Non-communicable diseases, 1848–2000 (Sources: HMSO; author's calculations)

Starting with about 50% each, the share of the non-communicable rose to about 80%, while that of the infectious ones shrank to 20% or less in the post-transition 'age of degenerative and man-made diseases.'

Meanwhile, the solid line in Fig. 2.8 is the level of non-communicable diseases. Their long-term path has two main features: (1) an upward trend during the second half of the nineteenth century; (2) the downward long-term trend over much of the twentieth century. This chapter will mainly focus on the downward segment, though both play a significant role in the book. The broken line is infectious diseases from Fig. 2.2, reproduced here to clarify the relative long-term paths of the *levels*:

Proposition 2 *In the Malthusian regime, the rates of non-communicable diseases may have been at least as large as infectious ones.*

Why 'at least as large?' The data reveal only one cause per person, not the accompanying causes. The reported deaths from infections, especially during outbreaks, may have been accompanied by degenerative conditions that likely went unrecorded; the opposite is possible as well where degenerative conditions may have been accompanied by various infections.

In another notable feature of Fig. 2.8, non-communicable diseases began falling about four decades after the infectious ones had started receding. The next proposition clarifies the paths of the two categories between the regimes:

Proposition 3 *In the Transition, while the fall of non-communicable diseases lagged that of the infectious ones, the overall long-term rates of both categories fell during the twentieth century.*

The data clear up the issue of levels versus shares of the non-communicable diseases: the trend in the *share* of non-communicable diseases was upward, but the trend in their *level* during the twentieth century was downward, for most of the century. The data make it plain that any perspective of the Transition without the non-communicable category would be partial at best: about half of what happened in the Malthusian regime and increasingly more since then would stay hidden. Since such diseases are linked to aging, our grasp of how aging has advanced since the nineteenth century may have gaps as well.

To narrow the gaps, more questions may need to be asked. For instance, in the Malthusian regime, average lives were likely shorter and the rates of non-communicable diseases heftier than in the post-transition era. What does that convey about the likely average age at which such diseases turned fatal? One may conjecture:

Proposition 4 *The burden of non-communicable diseases on average was heavier even at relatively younger ages in the Malthusian regime than in the post-transition regime.*

If so, deaths caused by them were likely being 'delayed' in the transition stage. Or, the 'age of delayed degenerative diseases,' the fourth stage added to the Transition since the 1980s, may not have been so recent. The next chapter explores when and among which adult-age groups the delays likely began before they were spotted in the middle-to-older age people relatively recently.

The downward trend in non-communicable diseases also seems paradoxical. As the population was aging more than ever before, the rates of such diseases should have risen over the twentieth century. The paradox is that as the population was aging, the rates of such diseases were falling instead. It can be stated another way: average lives were longer in the post-transition regime than in the Malthusian one; longer lives make people more susceptible to non-communicable diseases; one would then expect the rates of non-communicable diseases to have been heftier in the post-transition regime than in the Malthusian one. But the evidence shows just the reverse: it was the Malthusian stage with shorter average lives that had the heftier rates of such diseases.

Moreover, given finite human lifespan, the 'delays' imply the compression of deaths. How may the compression figure in the downward trend? Recall from last chapter that each aggregate rate coincides with a period-profile, which, in turn, is made up of contributions by many cohorts. Subsequent chapters relate the cohort profiles to the period-profiles, and ultimately to the compression and the downward trend.

One more feature germane to aging is that the long-term trend of both categories is downward, as if there were a synergy between them. If so, why did both categories not fall together? What kind of synergism would generate the three to four decade lag? This issue, too, is linked to cohort-level change, explored in Chap. 4. The main development later chapters uncover is that as the epidemiologic situation of successive cohorts of children was improving, their cohort-profiles of non-communicable diseases were becoming better, which in turn influenced the period-profiles to which those cohorts contributed until years later, shaping the long-term trend in the aggregate rates of non-communicable diseases. The synergism was likely intertemporal, and would take some time to show up in the long-term trends; hence the lag.

Meanwhile, Table 2.2 provides some clues about aging in the contributions of the two diseases-categories to the fall of the all-disease rate. Consider the segment under 'aggregate population.' The age-standardized death rates per 100,000 are averages for the time intervals in the first column. In each row the rates for the categories sum up to the all-disease rate. Between 1870/1874 and 1950/1954 the all-disease rate fell from 2,831.6 to 1,607.6, or by 1224, of which −945.2 was

Table 2.2 The contributions of the disease categories to the reduction of all-disease-caused death rate

	All diseases	Infectious diseases	Non-communicable diseases
	Aggregate population		
Average rate 1870/1874	2,831.6	1,181	1,650.6
Average rate 1950/1854	1,607.6	235.8	1,371.9
Average rate 1996/2000	1,045.4	180.0	865.4
Contribution through 1950/1954		77.2%	22.8%
Contribution through 1996/2000		56%	44%
	Age 0–14		
Average rate 1870/1874	402.1	255.7	146.4
Average rate 1950/1954	37.8	9.3	28.5
Average rate 1996/2000	8.5	0.8	7.7
Contribution through 1950/1954		67.7%	32.3%
Contribution through 1996/2000		64.8%	35.2%
	Age 15–64		
Average rate 1870/1874	918.7	341.2	423.2
Average rate 1950/1954	320.9	69.2	260.5
Average rate 1996/2000	170.1	2.2	155.9
Contribution through 1950/1954		72.8%	27.2%
Contribution through 1996/2000		64.3%	35.7%
	Age 65 and above		
Average rate 1870/1874	1,510.8	429.7	1,081.0
Average rate 1950/1954	1,248.9	166.1	1,082.8
Average rate 1996/2000	866.8	165.0	701.8
Contribution through 1950/1954		100.7%	−0.7%
Contribution through 1996/2000		41.1%	58.9%

Sources: HMSO; Human Mortality Database; author's calculations
Note: All rates are age-standardized per 100,000, using the 1994 standard weights

the contribution of infectious diseases (235.8–1,181), and −278.7 was that of non-communicable ones (1,371.9–1,650.6). Thus 77.2% of the total's decrease came from infectious diseases (−945/−1,224); and 22.8% from the other (−278.7/−1,224). Likewise, over the longer interval through 1996/2000, 56% of the fall in the all-disease rate came from infectious diseases and 44% from the non-communicable ones.

Figure 2.8 clarifies why the relative contributions in Table 2.2 shift when they are observed over the longer time interval. About 95% of the total fall in infectious diseases was achieved by 1950 and their marginal contribution has since been minimal. In contrast, only about a third of the total fall in the non-communicable was completed by then, making substantial progress since. Consequently, the total contribution of the infectious shrank from 77.2 to 56%, while that of the non-communicable rose from 22.8 to 44%.

Consider, further, the column for infectious disease, which shows that they were the main (proximate) contributor through 1950/1954 for all age groups, indicating that the containment of such diseases had involved the masses. As the larger contributor through 1950/1954, infectious diseases likely boosted life expectancy more than non-communicable diseases. The roles, however, reverse after 1950/1954. The following propositions summarize these features:

Proposition 5 *In the transition stage, the fall of infectious diseases accounted for the bulk of the fall in all-disease rate for the entire population and for the broader age groups.*

And, since the calculation of period-life expectancy uses data on current deaths, largely caused by disease:

Proposition 6 *In the transition stage, the fall of infectious diseases likely added more than that of non-communicable ones to period-life expectancies at various ages.*

Since the contributions changed as the interval expanded:

Proposition 7 *In the Transition, while both categories helped boost period-life expectancy, the non-communicable ones continued to help raise life expectancies on the margin after the transition stage.*

Moreover, for people 65 years or older, over 1870/1874–1950/1954, infectious diseases were the main contributor. The relative contributions may explain why life expectancy for 65-year olds barely grew through the first half of the twentieth century. Underneath the surface, infectious diseases in this age group were falling, but that was being more or less offset by the elevated rates of the non-communicable ones. However, that changed after the 1950s, when non-communicable diseases, too, began receding in older people: the 'age of delayed degenerative diseases.'

While the delays for older people may have been spotted recently, the delays, per se, may have been occurring for long. To see this, compare the contribution of non-communicable diseases in the age groups 15–64 and 65+ in the bottom half of

Table 2.2. Through 1950/1954, it was 27.2% for 15–64, but more or less zero for 65+. After 1950/1954, however, both are positive. The younger group experienced delays before and after 1950, while the older one did so mainly after then.

It indicates that there may have been cohort-like movements in the data. For instance, the 75-year-olds spearheading the new 'age of delayed degenerative diseases' among the older groups in 1975 might have been the 35-years old spearheading the delays among the youth in 1935. This relates to Proposition 4, which stated that such diseases may have been fatal at a much younger age in the past, and that the progress since then may have been like an age of delays as well, just occurring in different age groups. The reason why the delays were observed more prominently in the older age groups recently could be that a cohort-specific change that first shows up in the 35-year olds, for instance, would take four decades to show up in the 75-year olds. The focus on children's development in this book is motivated by just such a pattern. What occurred in 35-year olds in 1935, too, may have been swayed by what happened in their childhood years in the early 1900s.

In the second half of the nineteenth century, however, the feature relevant to aging is the upward trend of non-communicable diseases. The childhood years relevant to this horrible trend were in the 1820s through 1850s. The second half of the nineteenth century appears to have been an era of expansion: the rates growing and in progressively younger age groups. Moreover, the climbing rates coincided with growing period-life expectancies, mainly because infectious diseases were falling. The expectancies may thus be a deficient indicator of human development for this time, as the main trend of at least 50% of all diseases was running counter to what it was indicating.

In sum, the time-series view of the Transition indicates that cohort-like change may have been shaping the trends of non-communicable diseases and of aging. The two parameters α and β discussed in the previous chapter may help uncover the connections. While the focus in the rest of the book will be on cohort-level change, the time series trends will figure prominently throughout. The cohort parameters will be repeatedly tested for how well they help simulate the time series of non-communicable diseases.

References

Allen, R. (2009). Engels' pause: Technical change, capital accumulation, and inequality in the British industrial revolution. *Explorations in Economic History, 46*(4), 418–435.

Carnes, B. A., & Olshansky, J. (1993). Evolutionary perspectives on human senescence. *Population and Development Review, 19*(4), 793–806.

Easterlin, R. A. (1999). How beneficent is the market? A look at the modern history of mortality. *European Review of Economic History, 3*(3), 257–294.

Farr, W. (1885). In *Vital statistics: A memorial volume of selections from the reports and writings of William Farr,* Humphreys, republished Metuchen, 1975, New Jersey: Scarecrow Press.

Flinn, M. W. (1965). *Report of the sanitary conditions of the laboring population of Great Britain by Edwin Chadwick, 1842*. Edinburgh: Edinburgh University Press.

Floud, R., & Wachter, K. (1982). Poverty and physical stature: Evidence on the standard of living in London boys, 1770–1870. *Social Science History, 6*, 422–452.

Floud, R., Wachter, K., & Gregory, A. (1990). *Height, health, and history: Nutritional status in the United Kingdom, 1750–1980*. Cambridge: Cambridge University Press.

Floud, R., Fogel, R. W., Harris, B., & Hong, S. C. (2011). *The changing body, health nutrition and human development in the western world since 1700* (NBER). New York: Cambridge University Press.

Fogel, R. W. (1993). Economic growth, population theory and physiology: The bearing of long-term processes on the making of economic policy. *American Economic Review, 84*(3), 369–395.

Frazer, W. M. (1950). *History of English public health 1834–1939*. London: Bailliere, Tindall and Cox.

Fries, J. F. (1980). Aging, natural death and compression of morbidity. *New England Journal of Medicine, 303*(3), 130–135.

Fries, J. F. (1989). The compression of morbidity: Near or far? *Milbank Memorial Quarterly, 67*(2), 208–232.

Fries, J. F. (2005). The compression of morbidity. *Milbank Quarterly, 83*(4), 801–823.

Harris, B. (1994). Height, health and history: An overview of recent developments in anthropometric history. *Social History of Medicine, 7*, 297–320.

Harris, B. (2004a). *The origins of the British welfare state: Social welfare in England and Wales 1800–1945*. New York: Palgrave, Macmillan.

Harris, B. (2004b). Public health, nutrition and the decline of mortality: The Mckeown thesis revisited. *Social History of Medicine, 17*, 379–407.

HMSO. (1857). *The nomenclature of diseases drawn up by the joint committee appointed by the Royal College of Physicians of London* (1st, 2nd and 3rd ed.). London: HMSO.

Horrel, S., & Oxley, D. (1999). Crust or crumb? Intra-household resource allocation and male breadwinning in Late Victorian Britain. *Economic History Review, 52*, 494–522.

Humphries, J. (2013). *Childhood and child labor in the British industrial revolution*. Cambridge: Cambridge University Press.

Kohn, G. C. (1998). *Encyclopedia of plague and pestilence*. Hertfordshire: Wordsworth Reference, Wordsworth Editions Ltd.

Mckeown, T. (1976). *The modern rise of population*. London: Edward Arnold.

Mercer, A. (1990). *Disease mortality and population in transition*. Leicester: Leicester University Press.

Oeppen, J., & Vaupel, J. (2002). Broken limits to life expectancy. *Science, 296*, 1029–1031.

Olshansky, S. J., & Ault, A. B. (1986). The fourth stage of the epidemiologic transition: The age of delayed degenerative diseases. *Milbank Memorial Fund Quarterly, 64*(3), 355–391.

Omran, A. R. (1971). The epidemiologic transition: A theory of epidemiology of population change. *The Milbank Quarterly, 49*(4), 509–538.

Omran, A. R. (1982). Epidemiologic transition. In J. A. Ross (Ed.), *International encyclopedia of population* (pp. 172–183). London: The Free Press.

ONS. (1997). *Health of adult Britain 1841–1994*, Vol. I and II (Decennial Supplement no. 12 and 13), edited by J. Charlton, & M. Murphy. London.

Oxley, D., & Meredith, D. (2014). Nutrition and health, 1700–1870. In R. Floud & J. Humphries (Eds.), *The Cambridge economic history of modern history volume 1*. Cambridge: Cambridge University Press.

Pelling, M. (1978). *Cholera, fever and english medicine, 1825–1865*. Oxford: Oxford University Press.

Preston, S. H., & Haines, M. R. (1991). *Fatal years: Child mortality in late-19th century America* (Vol. 151, pp. 276–309). Princeton: Princeton University Press.

Rosenbaum, S. (1988). 100 years of heights and weights. *Journal of the Royal Statistical Society, 151*, 276–309.

Sachs, J. (2014). *Responding to Ebola*. http://www.project-syndicate.org

Sen, A. K. (1981). *Poverty and famines: An essay in entitlement and deprivation*. Oxford: Oxford University Press.

Szreter, S. (1988). The importance of social intervention in Britain's mortality decline circa 1850–1914: A re-interpretation of the role of public health. *Social History of Medicine, 1*(1), 1–37.

Szreter, S. (1997). Economic growth, disruption, deprivation, disease, and death: On the importance of politics of public health for development. *Population and Development Review, 23*(4), 693–728.

Szreter, S. (2005). *Health and wealth: Studies in history and policy*. Rochester: University of Rochester Press.

Szreter, S., & Mooney, G. (1998). Urbanization, mortality, and the standard of living debate: New estimates of the expectation of life at birth in the nineteenth century British cities. *Economic History Review, 51*(1), 84–112.

Tanner, J. M. (1990). *Fetus into man: Physical growth from conception to maturity*. Cambridge, MA: Harvard University Press.

Wohl, A. S. (1983). *Endangered lives: Public health in Victorian Britain*. Cambridge, MA: Harvard University Press.

Woods, R. (1985). The effects of population redistribution on the level of mortality in nineteenth century England and Wales. *Journal of Economic History, 45*(3), 645–651.

World Health Organization. (1978). *Manual of international statistical classification of diseases, injuries and causes of death: Ninth revision*. Geneva: WHO Press.

Wrigley, E. A., & Schofield, R. S. (1981). *The population history of England, 1541–1871: A reconstruction*. Cambridge, MA: Harvard University Press.

Chapter 3
The Profiles of Aging

3.1 Across Generations

The generation method helps study cohort-like movements. Used in the 1950s to study trends in the rates of tuberculosis (e.g. Springett 1952; Spicer 1954) the method has roots in the 1930s, when scholars noted that death rates in people born at different times appeared as if different generations were aging along unique paths (Kermack et al. 1934; Fridlizius 1989; Harris 2001).

A simplified example may clarify the method: the rates reported for the 25-year-olds in 1925, the 35-year-olds in 1935, the 45-year-olds in 1945, the 55-year-olds in 1955, the 65-year-olds in 1965, and the 75-year-olds in 1975, all pertain to people born in 1900. These rates form an age-profile of the people born in 1900. Such a profile can be made for people born in 1800, 1810, 1820, 1830, 1840, 1850, 1860, and 1870, through 1940, as if they were a series of generations. The ones born in 1800, 1810, 1820, 1830, 1840, 1850, and 1860 were raised in the Malthusian stage; the ones in 1870 through 1920 grew up in the transition stage; and the rest in the post-transition one. Their profiles can be compared to study how aging had varied across them.

Over 1800–1940, however, there are just 15 profiles and that is not enough to ensure robust inference. Since no person can die twice, the hurdle can be overcome by selecting another series of profiles for a different set of birth-years. For instance, the set of birth years 1805, 1815, 1825, 1835, 1845 through 1945 are 15 more profiles pertaining to different people. Here, for instance, the 25-year-olds in 1830, 35-year-olds in 1840, 45-year olds in 1850, 55-year-olds in 1860, 65-year-olds in 1870 and 75-year olds in 1880, were all born in the year 1805. The profiles of the birth-years 1805, 1815, 1825, 1835, 1845, 1855 and 1865 pertain to people raised in the Malthusian regime; the ones for 1875 through 1925 in the transition stage; and the rest in the post-transition stage. Between the two sets there are now 30 profiles: about 12 Malthusian, 14 transition stage, and 4 post-transition.

S. Arora, *The Transitions of Aging*, International Perspectives on Aging 12,
DOI 10.1007/978-3-319-14403-0_3

Table 3.1 Sets of age-profiles by decade of birth

Birth decades					
Regime	I	II	III	IV	Average
Malthusian		1795–1804		1796–1805	1795–1805
Malthusian	1800–1809	1805–1814	1801–1810	1806–1815	1800–1815
Malthusian	1810–1819	1815–1824	1811–1820	1816–1825	1810–1825
Malthusian	1820–1829	1825–1834	1821–1830	1826–1835	1820–1835
Malthusian	1830–1839	1835–1844	1831–1840	1836–1845	1830–1845
Malthusian*	1840–1849	1845–1854	1841–1850	1846–1855	1840–1855
Malthusian*	1850–1859	1855–1864	1851–1860	1856–1865	1850–1865
Early transition	1860–1869	1865–1874	1861–1870	1866–1875	1860–1875
Early transition*	1870–1879	1875–1884	1871–1880	1876–1885	1870–1885
Early transition*	1880–1889	1885–1894	1881–1890	1886–1895	1880–1895
Mid-transition	1890–1899	1895–1904	1891–1900	1896–1905	1890–1905
Mid-transition	1900–1909	1905–1914	1901–1910	1906–1915	1900–1915
Late transition*	1910–1919	1915–1924	1911–1920	1916–1925	1910–1925
Late transition*	1920–1929	1925–1934	1921–1930	1926–1935	1920–1935
Post-transition	1930–1939	1935–1944	1931–1940	1936–1945	1930–1945
Post-transition*	1940–1949	1945–1954	1941–1950	1946–1955	1940–1955
Observations	15	16	15	16	

With that basic design in mind, now consider how the data are actually reported, which is slightly more complicated, but nearly as pliable: the rates in any given year are reported for age groups such as 25–34, 35–44, 45–54, 55–64 and 65–74. Instead of pertaining to a particular year of birth, they pertain to a decade of birth. For example, the 25–34-year olds in 1934, the 35–44-year olds in 1944, the 45–54-year olds in 1954, the 55–64-year olds in 1964, and the 65–74-year olds in 1974, were all born sometime in 1900–1909. The birth decades for this set are the 1810s, 1820s, 1830s, 1840s, 1850s, 1860s, and 1870s through the 1940s. Add another 15 profiles from the next batch of birth decades, 1805–1814, 1815–1824, 1825–1834, 1845–1854, 1855–1864 through 1945–1954, and the two sets amount to about 30 age-profiles.

Following the strategy, the book isolates four sets of profiles listed in Table 3.1. They each pertain to different people even though the birth decades across them are similar. The proximity of the birth-decades in sets I and III could mean that the results for them can be close though not identical; as could the results for sets II and IV. The last column states the birth-years to which the average result across the sets might pertain. For analysis, moreover, the profiles in sets I and II are staggered enough in time to be seen as a series of 31 observations; likewise for sets III and IV. Sorted by the middle year of their birth-decades, they can amount to 62 observations for the α and β of the profiles.

Further, there are a couple of unfinished profiles in each set. The working assumption is that the 65–74-year-olds and the 55–64-year olds in the 1940s profiles had fared no better than the 65–74-year olds from the 1920s and the

55–64-year olds from the 1930s profiles. At the other end, the 25–34-year olds in the first two profiles, and the 35–44-year olds in second one fared no worse than the ones born in the 1801s and 1810s.

Finally, why do the profiles have 10-year age groups limited only to age 25–74 of the adult life cycle? Since people become increasingly prone to non-communicable diseases as they age over those 50 years, it is an apt period for studying the progress of aging. Moreover, apples-to-apples comparison requires that the age groups be set uniformly across the profiles. Only for the groups between ages 25 and 74 can uniformity be achieved as the format of the reported data has changed at least thrice over the past 150 years. Most adult groups were originally reported in 10-year age intervals, the last five reported as 45–54, 55–64, 65–74, 75–84 and 85+; that alone dictates 10-year groups for the entire period, but it leaves open the issue of the oldest age group that could be included. After minor change in the 1901s, the format was 5-year intervals in the 1920s, the last four reported as 65–69, 70–74, 75–79, 80+, which later became more granular 80–84 and 85+. The varied formats leave only 65–74 and 75+ as the oldest that can be set uniformly over the 150 years. The 75+ does not match up to unique birth decades; hence, the profiles run from age 25–34 through 65–74.

The next two sections first study cross-regime change in profiles 'representative' of each regime. Except the post-transition profile, the representative ones are the arithmetic averages of the profiles marked by asterisks in Table 3.1. The last section studies the trends in successive profiles since the nineteenth century.

3.2 Expose Key Features of Aging

Figures 3.1 and 3.2 show the cross-regime change in the profiles from sets I and II. Their first notable feature is that all of them slope upward exponentially, affirming that the rates of non-communicable diseases climb with age. The biological basis for this upward slope is said to be senescence, a non-inheritable trait of internal degeneration that becomes increasingly potent with age and ensures finite lifespan (Medawar 1952; Williams 1957; Kirkwood and Holliday 1979, 1981; Hayflick 1985; Kirkwood 1985, 1999; Finch 1990; Rose 1991). It is critical to keep in mind that the profiles are not senescence itself. Seen as a non-specific biological force, it is not directly linked to any code in the International Classification of Diseases. The slopes only affirm that senescence likely intensifies with age and that the rates of dysfunctions or diseases related to aging tend to climb as cohorts become older.

Second, the profiles had shifted. At each age, the Malthusian one had heftier rates than others. But despite the shifts, they stayed upward-sloping.

Third, though the profiles are distinct, they tend to be converging at older ages. Clearer in Fig. 3.2, the convergent tendency seems to suggest that there is a limit to lifespan, even if a unique one might be unknown.

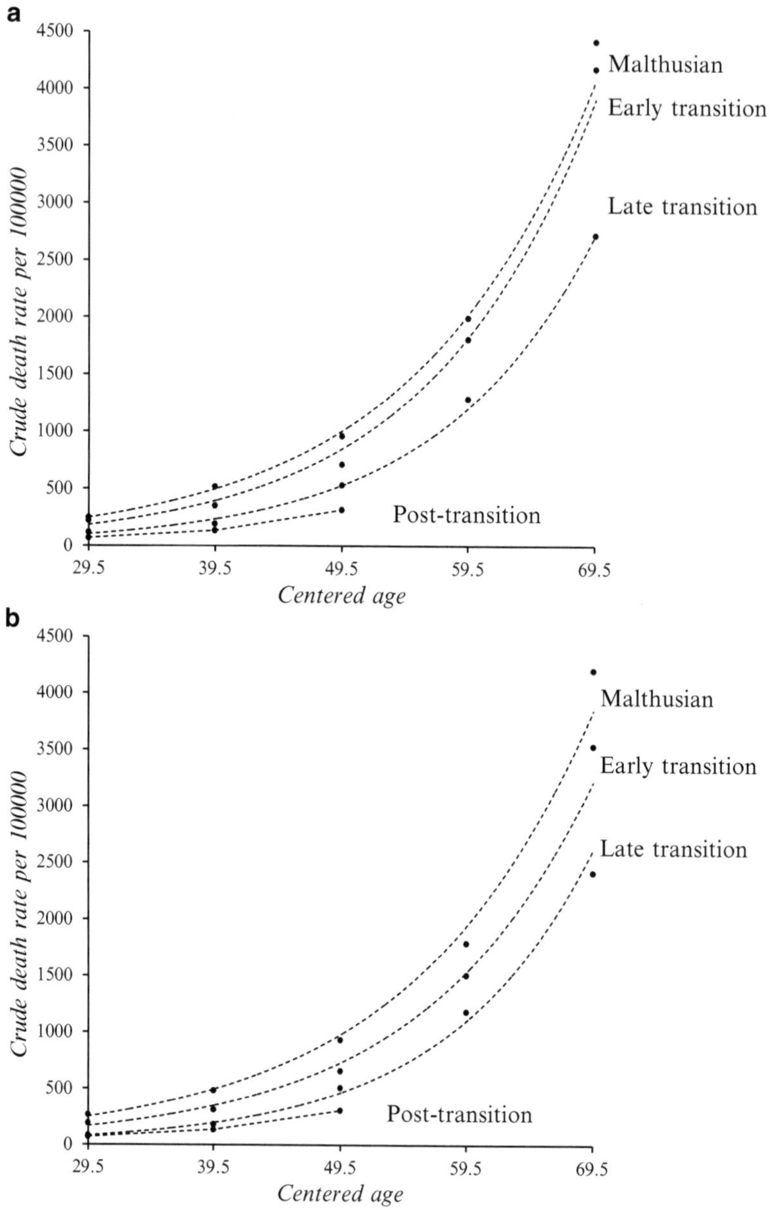

Fig. 3.1 The exponential trend in the age-profiles of non-communicable diseases across regimes (**a**) Set I (**b**) Set II (Sources: HMSO; author's calculations)

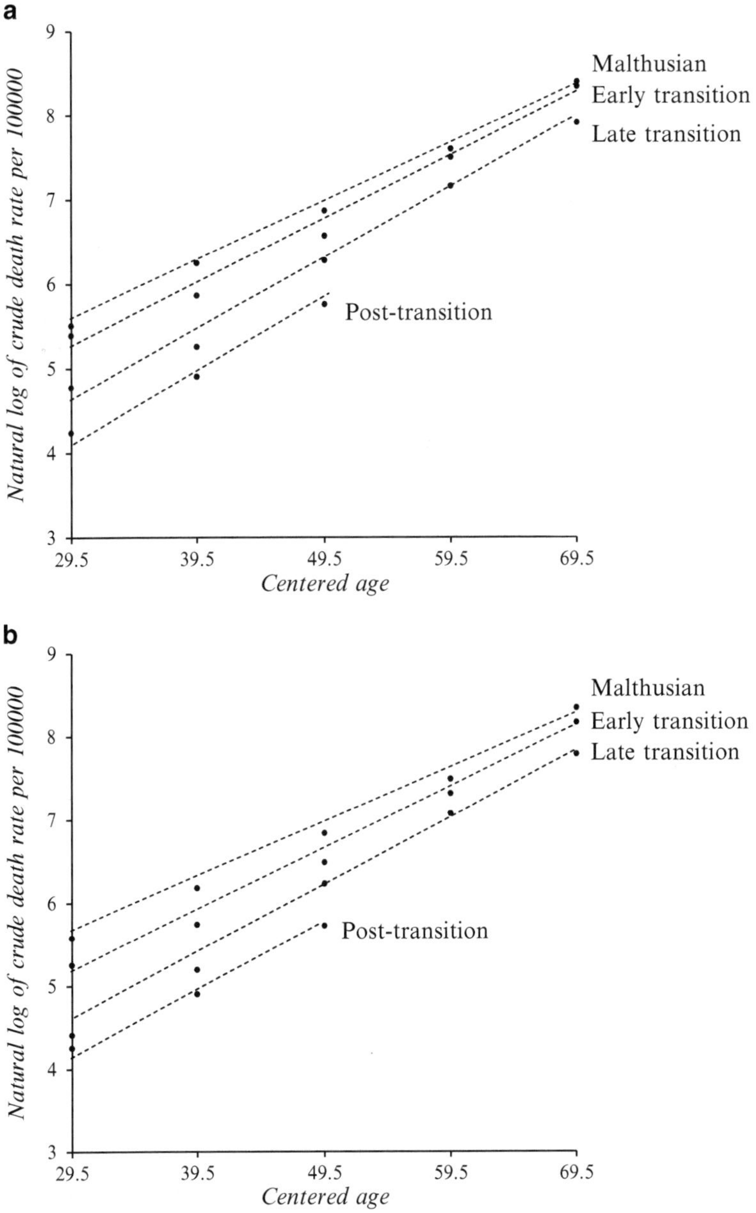

Fig. 3.2 Natural log of the age-profiles in Fig. 2.1 (**a**) Set I (**b**) Set II (Sources: HMSO; author's calculations)

Fourth, each profile starts at a different level. Visible more clearly in Fig. 3.2, the Malthusian one had the worst start and the post-transition profile the best. They indicate that change across generations can show up early in the adult years.

Fifth, the slopes of the newer profiles were progressively steeper. This type of progress is similar to the tendency toward "rectangularization" in period-survival curves. Facing some likely maximal age, fewer deaths at younger ages in the newer profiles also means that the rates must eventually speed up as more people come ever closer to the limit.

In a near-ideal theoretical case, most everyone in a profile would live until just before the maximal age, and then death rates would rise sharply. How might the ideal shed light on what may be going on in Fig. 3.2? The eventual sharp rise of the death rates means that the ideal profile will also be the steepest. However, most of its members will have lived the longest life possible, or will have potentially aged the most. The steepest profile, therefore, must also be the one that displays the most aging potential. In the progress toward such an ideal, successive profiles would be shifting down and becoming steeper; and would be moving away from profiles with less aging potential toward ones with more potential. That is just what Fig. 3.2 seems to convey: the aging potential across the profiles had grown; the Malthusian one had the least potential.

Those telling features about levels, slopes and aging potentials pose the practical question of how to quantify them. Following a line of inquiry since at least Gompertz (1825, 1872), biodemographers and biologists state the age-profile as an exponential curve.

$$M(c, a) = \alpha e^{\beta a}. \tag{3.1a}$$

Here, $M(c, a)$ is the 'force of mortality' for cohort c, at age a, reckoned best by the crude death rate at that age. The parameter α is the initial state, and β expresses the upward or the positive slope. This is identical to the set up for the spending profile illustrated in the first chapter, except these are cohort-profiles not period-profiles. The next chapter relates the parameters of the cohort-profiles to the period-profiles, and eventually, Chap. 10 relates both to the profiles of healthcare spending.

Meanwhile, the issue is how to numerically express the aging potential in a profile. A measure called the 'mortality rate doubling time' helps interpret it (Kirkwood 1985; Finch 1990). Calculated as the natural logarithm of 2 (or about 0.693) divided by β, the doubling time expresses how quickly the rates of non-communicable diseases double with age. Usually, between two distinct profiles, all else held constant, the one with a smaller α and a larger β has the steeper slope, shorter doubling time, and thus better aging potential.

But why does shorter doubling time usually signify better potential? Dying more rapidly, but better aging potential? It helps to think about this issue in terms of that near-ideal profile, where most people end up living the longest possible lifespan before dying quickly as they reach the maximal age. That profile would have the smallest α and the steepest slope or β. With its β the largest, its doubling time would be the shortest. Downward shifts and steeper slopes signify that deaths are being deferred to older ages than before, as if the profiles are progressing toward the theoretical ideal. Thus, together with falling α, the shrinking doubling times correctly signify growing aging potentials.

Note also that the slope or the doubling time express the shape of the entire profile, enabling cross-profile comparisons, a feature suited to an aggregate-level inquiry, as opposed to a micro-level one in which, say, only the middle-to-older aged people across cohorts are compared.

Further, a vital issue Figs. 3.1 and 3.2 convey is that aging along a profile, or within a profile, has been true in all regimes. In any given profile, there were always people who had aged more than others, or had lived longer than others. What seems to set modern aging apart, however, is that entire profiles had shifted: a greater percentage of newer generations were able to reach each age. This cross-profile change, not the within-profile one, appears to have been the main thrust of aging during the Transition. Without it, the Malthusian-type of profile would be the only one left.

Consequently, the measure of aging suitable for such a situation would be the change in the β_i across the profiles. The within-profile aging is for a static profile, which has just one unchanging β. So, if an outlook of aging calls for a static profile, then it also calls for no change in the β, which also means that it expects the physiological underpinnings will not change whatsoever across generations. The brief history of modern aging, as more evidence will show in the next few sections, does not support such a premise. The book returns, over and again to this important matter, as the static profile seems to have been a pervasive assumption in most analyses of how aging relates to healthcare spending over time. Such a profile is at the root of the paradox of aging mentioned in the last chapter. This will be illustrated in Chap. 4. It also seems to be the basis for the notion that healthcare spending will soar in the years ahead, discussed in Chap. 9.

While comparing profiles, however, inference can potentially go astray. Consider, for instance, the convergent profiles in Fig. 3.2. The slopes of the ones converging from *below* are steeper; and their aging potentials are better. Steeper slope, shorter doubling time, and better aging potential correspond in a clear-cut way as shown in Fig. 3.3a for two stylized profiles I and II.

However, they do not correspond as clearly in an exceptional case in which the slope becomes flatter and the doubling time grows. The logical inference from the flatter slope alone is that the potential has worsened. But the inference is incorrect. This case brings out the importance of inspecting α before making an inference: if it has shrunk, then a flatter slope compared to another profile indicates gain in aging potential, as shown in Fig. 3.3b: death rates at each age in II were less than in I. Such a scenario is possible if there were a one-time structural change, with subsequent profiles becoming steeper and restoring the usual pattern of falling α_i and rising β_i. Further, the one-time change, too, may not be a global aggregate-level outcome. It is possible the profile may have not been observed for a long enough time or until an old enough age when, all else equal, the slope must become steeper.

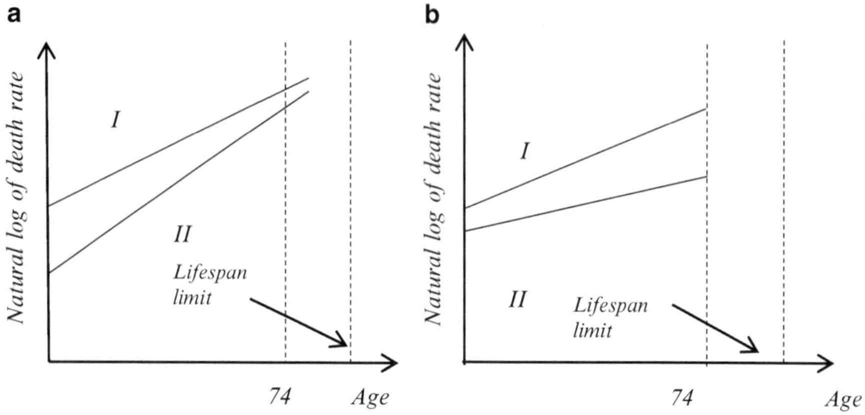

Fig. 3.3 Inferences from stylized age-profiles as the initial state changes (**a**) Converging slopes (**b**) Diverging slopes

Table 3.2 Inferences about aging from the age-profile parameters

For comparison of two hypothetical age-profiles I and II	
$\alpha_{II} < \alpha_I$	• Better initial state for II
$\beta_I < \beta_{II}$	• Steeper slope for II
	• Shorter doubling time for II
	• Better aging potential of II, if $\alpha_{II} < \alpha_I$
$\alpha_{II} < \alpha_I$ and $\beta_{II} < \beta_I$	• Worse aging potential of I; β_i and the doubling time alone can mislead, however; check α for correct inference
For time series of the parameters from successive age-profiles	
Initial states	• Falling α_i: improving initial states for the profiles • Growing α_i: worsening initial states for the profiles
Slopes	• Growing β_i: shrinking doubling times and growing aging potentials as long as α_i fall • Falling β_i: growing doubling times and falling aging potentials as long as α_i rise

For ages 25 through 74 studied in the book, all cross-regime comparisons for the aggregate show growing potentials like the ones in Fig. 3.3a. In the sub-categories of non-communicable diseases, however, there are a couple of instances like Fig. 3.3b, where the doubling time, taken at face value, misleads. They will be flagged to enable proper inference in subsequent chapters.

The top half of Table 3.2 sums up the main inferences for comparisons of the parameters of two profiles. The lower half extends the reasoning to a series of profiles: as long as α_i are falling, a downward trend in doubling times (or rising β_i) signifies growing aging potentials. An upward trend in doubling times (or falling β_i) along with rising α_i implies worsening aging potentials.

The fact that α_i and β_i usually relate inversely has been a cornerstone of studies of demographic aging across human populations, compared at a point in time (e.g. Strehler and Mildvan 1960). Though the framework is being used here to study aging across cohorts, from within a population over time, all empirical results in the book reaffirm the inverse long-term relation. Here, their long-term *trends* relate inversely. However, the inverse relation is not necessarily a mechanical one. That is, β_i grows as α_i shrinks is not a mere consequence of finite lifespan. The empirical results in Chap. 8 indicate that though there is some mechanical relation, α_i and β_i seem to have been swayed by socio-economic and political changes since the nineteenth century.

With the inferences in mind, consider Table 3.3 that shows how the α_i and β_i for the age-profiles in sets I through IV measure up. The top segment has arithmetic means of the four sets of results below it. The parameters were estimated by first re-writing Eq. 3.1a as

$$M(c, a) = \alpha e^{\beta a} \mu \tag{3.1b}$$

Table 3.3 Age-profile parameters of non-communicable diseases across regimes

Set	Regime	$\widehat{\alpha}$	$\widehat{\beta}$	R^2	Doubling time (years)
Mean	Malthusian	32.4	0.069	0.997	10.1
	Early transition	18.9	0.075	0.990	9.3
	Late transition	7.8	0.084	0.995	8.3
	Post-transition	4.5	0.088	0.994	7.9
I	Malthusian	31.3	0.0700	0.999	9.9
	Early transition	19.3	0.0764	0.985	9.1
	Late transition	9.3	0.0816	0.991	8.5
	Post-transition	4.1	0.0906	0.993	7.7
II	Malthusian	33.3	0.0683	0.995	10.2
	Early transition	18.7	0.0740	0.990	9.4
	Late transition	6.5	0.0863	0.997	8.0
	Post-transition	4.9	0.0858	0.993	8.1
III	Malthusian	30.3	0.0697	0.996	9.9
	Early transition	21.3	0.0722	0.989	9.6
	Late transition	8.2	0.0836	0.994	8.3
	Post-transition	4.3	0.0895	0.994	7.7
IV	Malthusian	34.6	0.0671	0.996	10.3
	Early transition	16.3	0.0767	0.995	9.0
	Late transition	7.0	0.0846	0.997	8.2
	Post-transition	4.8	0.0855	0.996	8.1

Note: The parameter estimates are for Eq. 3.2 in the text; $\widehat{\alpha}$ is the anti-log of the estimate of A; doubling time is natural log of 2, or 0.693, divided by $\widehat{\beta}$

where μ is a multiplicative or proportional error term; by taking natural logarithms on both sides of Eq. 3.1b; and finally by estimating:

$$m(c,a) = A + \beta a + \varepsilon, \qquad (3.2)$$

where, $m(c,a)$ is the natural logarithm of the crude death rate $M(c,a)$; A is natural logarithm of α; the ages 'a' are 29.5, 39.5, 49.5, 59.5 and 69.5, the mid-points of the age intervals; ε, the natural logarithm of μ, is a normally-distributed error term; and the estimation method is Ordinary Least Squares. In Table 3.3, and in all subsequent Tables, $\widehat{\alpha}$ are the anti-log of the estimates of A.

The fifth column indicates that exponential curves mimic the paths of the profiles quite well. Further, the results for sets I and III are close; as are the results for sets II and IV. The $\widehat{\alpha}_i$ convey that there was sizable change in the initial states: the Malthusian profile's $\widehat{\alpha}$ was about 1.7 times that of the early-transition profile; about 4.2 times that of the late transition profile, and about 7.2 times that of the post-transition one.

Further, as $\widehat{\alpha}_i$ fell, the slope parameter β_i grew from about 0.07 for the Malthusian profile to 0.09 for the post-transition one. The doubling time, as a result, shrank from about 10 years for the Malthusian, to roughly 8 years for the post-transition profile. The Malthusian profile had the heftiest burden at each age, and thus the poorest aging potential. Thus, unlike the scenarios of spending profiles presented in Chap. 1 (OECD 2013), for the rates of non-communicable diseases, smaller the β_i, flatter the slope, worse the profile. This is unlikely to be an underpinning for healthy aging.

What is the consequence of the different $\widehat{\alpha}_i$ and $\widehat{\beta}_i$? Say you want to compare two cohorts of equal size before the onset of adult years; and also say non-communicable diseases were the only reason for death. Subject one of the cohorts to the parameters of the Malthusian profile, and the other to the parameters of the post-transition profile. The cohort subject to the post-transition parameters would have 7.5% more people by about age 80 and nearly 13% more by age 90. If one cohort were already 10% smaller to begin with, then, subject to the Malthusian profile's parameters, it would have 17% fewer people by age 80, and about 10% fewer by age 90, than the other. Thus, all else equal, a profile with a steeper slope or larger β delivers more older-aged people than the other.

The following four propositions sum up the main features of the profiles and what they suggest about aging across the regimes. First, about their shapes:

Proposition 8 *The age-profiles of non-communicable diseases slope upward and climb exponentially with age.*

Second, about their locations and about the nature of aging:

Proposition 9 *While the upward slopes of all age-profiles signify heterogeneity of aging within them, their downward shifts signify growing aging potentials across them.*

Without the downward shifts, all aging would occur roughly along that one Malthusian profile, and the aggregate rates would have risen as increasingly more people had reached older ages. But the aggregate rates have fallen over the twentieth century, indicating that the downward shifts may have been shaping the falling long-term trend. By using the parameters of successive profiles, Chap. 4 demonstrates how the shifting profiles account for the long-term trends of the aggregate, including the upward trend in the second half of the nineteenth century.

Meanwhile, as each cohort's age advanced, the newer profiles were steeper, and had tended to converge toward with older ones. Hence, on the nature of the shifts:

Proposition 10 *The profiles shifted in a non-parallel, but convergent manner, and the gaps between them taper with age as all of them likely face some limit to the lifespan.*

The raw data give a better sense of the narrowing gaps. The rates in the Malthusian profile at age 35–44 were four times, and at age 45–54, three times the rates in the post-transition profile. Compared to the late transition profile, the rates in the Malthusian profile were 2.7 times at age 35–44, tapering to 1.5 times by age 65–74.

There are some telling cross-age differences as well: the rate for the Malthusian 35–44-year olds was 1.8 times that of the post-transition 45–54-year olds. The rates in even as young as the 25–34-year olds in the Malthusian profile were nearly as high (0.8–0.9 times) those of the 45–54-year olds in the post-transition profile. The cross-age comparison might echo the refrain "50 is the new 30 or 40," usually invoked to express that newer generations are faring better than the ones before. However, given the convergent nature of the shifts, at least for death rates, the refrain does not apply equivalently to older ages. For instance, "50 is the new 30 or 40" is not equivalent to "70 is the new 60 or 50." Meanwhile, since the profiles shifted early in the adult life cycle, and the stage of life before then is primarily the childhood years, one may conjecture:

Proposition 11 *Changes in children's development in the various regimes of the Transition could have influenced the shifts in the age-profiles.*

Chapter 8 will explore this issue empirically.

3.3 And Clarify the Debate About Its Prospects

The shifts help understand the 'age of delayed degenerative diseases.' Figures 3.1 and 3.2 clarify that the 'delays' first surfaced in the young and then at older ages as the decades passed. For instance, compared to the 35-year olds of the 1890 profile, a step-up for the 35-year-olds from the mid-transition profile (1900) would show up in 1935. It would then take another four decades for any difference to show up between their 75-year olds in 1965 and in 1975. Successive profiles going through similar change at younger ages in the first half of the nineteenth century would decades later appear as a seemingly-new trend in the middle-to-older age groups – or, as the age of delayed degenerative diseases. In that sense, much of the twentieth century, and not just its last few decades, seems to have been the age of delays.

It helps to recap the events leading up to the debate on aging in the 1970s and 1980s. The evidence through the first half of the twentieth century seemed to back the view that period-life expectancies at older ages may not grow as the rates of non-communicable diseases had stayed relatively elevated in the 'age of degenerative and man-made disease.' Period-life expectancy at, say, age 65, had inched up barely two years during the first half of the twentieth century. Since most non-communicable diseases climb with age, their rates were expected to rise further as more people would reach middle-to-older age than before.

However, when increasingly more people started reaching those ages in the 1960s and 1970s, the rates began falling instead of climbing. Period-life expectancies came in better than the presumed limit. That triggered the debate on whether the new trends were transient or permanent; whether period-life expectancies would run into some new limit or would grow continually.

The case for transience said that the deaths in this new 'age of delayed degenerative diseases' were merely being delayed (Lee and Carter 1992; Olshansky et al. 1990; Horiuchi 1989; Olshansky 1985; Olshansky and Ault 1986). People would in the end reach even older age when rates of such diseases grow. Citing upward slopes of period-profiles, it argued that there must be some limit to lifespans even if one could not be specified a priori. So there should be some limit to life expectancies as well.

Figures 3.1 and 3.2 affirm that all profiles slope upward, and agree that there must be some maximal, but unknown, age limit. However, they also convey that instead of one universal upward-sloping profile, there are several unique ones. Despite their positive slopes, greater fractions of the newer cohorts may reach older age. Period-life expectancies could grow when people in those newer profiles reach an advanced age.

An upward slope, by itself, thus does not ensure that life expectancies may stall. The downward shifts have to stop as well. Only when that happens would there be roughly one universal profile that all generations would chart.

Meanwhile, the case for permanence said that there appears no a priori limit to period-life expectancies (Vaupel and Lundström 1994; Oeppen and Vaupel 2002; Rau et al. 2008). Citing long-term trends in even in the oldest-of the old such as centenarians (e.g. Thatcher et al. 1998), it argued that setting new limits could turn out to be just as illusory as the erstwhile ones.

While the shifted profiles in Fig. 3.2 and the evidence in Table 3.3 support such a case, they offer reason for caution as well: the shifts have not been parallel. Their convergent nature indicates that the gains shrink as age advances. While the percentage change may be sizable, the absolute change may not be large.

Either side of the debate could find support in the cross-profile evidence, but only from one aspect of it. The cautious case may be argued on the basis of upward slopes; the other may be claimed on the basis of the shifts. The complete evidence, however, calls for a synthesis: all profiles slope upward, but they shift in a non-parallel and convergent way. It is important to note that in the synthesis the reason to be cautious about the outlook is not the upward slope, per se; all profiles have that trait. The reason is their convergent tendency. Period-life expectancies could still grow, but the gains at progressively older ages may not be equivalent to the ones at younger ages.

Moreover, the shifts did not pertain to just the middle-to-older age segments. They began much earlier in the adult life cycle and the profiles traced unique paths. Thus, studying age-profiles since early on in the life cycle, before they tell on the trends in the very old-aged people, may help frame the likelier outlooks for aging.

One such outlook, for example, is the compression of deaths (Fries 1980, 1989). Though its estimate of the maximal lifespan may have turned out to be off the mark, the framework may have been correct in assuming that there must be a limit. If so, deaths can be delayed only so far, and would be increasingly confined to the years just before the limit. That is, if the limiting age—which is not the same thing as life expectancy—is, say, 100 years, then people would die increasingly over some shrinking interval before age 100.

The reader may have realized that compression is much like the gradual progress toward "rectangularization"—or toward that near-ideal theoretical case discussed earlier. It is about shrinking doubling times, or about growing aging potentials across the profiles: they shift down as death rates at young adult ages reduce; but as the deaths are delayed to older ages the profiles also become steeper. Thus, the cross-cohort evidence expresses compression in terms of profiles shifting down, becoming steeper, and converging at older ages.

However, the profiles presented here and the ones usually cited in the debate differ in a basic way. The debate refers to period-profiles. By using the data in period-profiles, period-life tables calculate the probabilities of death at various ages, converting them into period-life expectancies stated in terms of years. Shifting period-profiles mean changing period-life expectancies. Since many cohorts contribute to the shifts in the period-profiles, shifting cohort profiles may sway period-life expectancies as well.

Moreover, an outlook of period-life expectancy is based on what the period-profile may look like in the future. Studying how the period-profiles in the past may have been influenced by cohort-level change could help inform how such an outlook could be framed. The next chapter studies the matter in some detail and revisits the debate. Meanwhile, the next section studies if some sub-categories of non-communicable diseases had echoed the cross-regime pattern of the aggregate.

3.4 As Do the Sub-categories of Non-communicable Diseases

What could the profiles of various sub-categories reveal that the ones for the aggregate do not already show? The non-specific nature of senescence implies that multiple organs of the body can impair with age. For example, damaged organs of the nervous system may have accompanied deaths reported from some circulatory disease. However, the data reveal only one cause per death (e.g. circulatory), not the accompanying one (e.g. nervous). One way to check if both types followed the broad pattern of the aggregate is to study the reported nervous system diseases and the circulatory ones separately. The purpose is neither to dismiss the extra burden of co-morbidity nor to estimate it. It is to

Table 3.4 Age-profile parameters of the sub-categories of non-communicable diseases, average across sets I, II, III and IV

Regime	$\widehat{\alpha}$	$\widehat{\beta}$	R^2	Doubling time (years)
	Circulatory system			
Malthusian	7.4	0.0795	0.995	8.7
Early transition	2.7	0.0942	0.981	7.3
Late transition	1.1	0.1039	0.990	6.7
Post-transition	0.4	0.1146	0.999	6.1
	Nervous system			
Malthusian	9.2	0.0344	0.957	20.2
Early transition*	6.4	0.0255	0.966	27.5
Late transition	1.0	0.0511	0.898	13.7
Post-transition	0.5	0.0614	0.980	11.3
	Digestive system			
Malthusian	9.7	0.0437	0.973	15.9
Early transition	5.8	0.0455	0.987	15.2
Late transition	0.6	0.0723	0.974	9.7
Post-transition	0.2	0.0902	0.980	7.7
	Musculoskeletal system			
Malthusian	0.8	0.0579	0.844	12.0
Early transition	0.2	0.0638	0.978	10.9
Late transition	0.04	0.0879	0.857	8.0
Post-transition	0.03	0.0840	0.940	8.3
	Neoplasms*			
Malthusian	1.3	0.0975	0.968	7.1
Early transition	1.6	0.0931	0.980	7.5
Late transition	1.4	0.0969	0.990	7.1
Post-transition	0.9	0.1003	0.998	6.9

Note: The parameter estimates are for Eq. 3.2 in the text; $\widehat{\alpha}$ is the anti-log of the estimate of A; doubling time is natural log of 2, or 0.693, divided by $\widehat{\beta}$

Table 3.5 Age-profile parameters of the sub-categories of non-communicable diseases, set I

Regime	$\widehat{\alpha}$	$\widehat{\beta}$	R^2	Doubling time (years)
	Circulatory system			
Malthusian	8.2	0.0781	0.998	8.9
Early transition	2.8	0.0932	0.982	7.4
Late transition	1.4	0.1017	0.993	6.8
Post-transition	0.3	0.1174	0.999	5.9
	Nervous system			
Malthusian	9.8	0.0359	0.965	19.3
Early transition*	6.3	0.0260	0.989	26.7
Late transition	1.4	0.0457	0.855	15.2
Post-transition	0.5	0.0613	0.967	11.3
	Digestive system			
Malthusian	10.7	0.0438	0.982	15.8
Early transition	6.2	0.0452	0.991	15.3
Late transition	0.8	0.0649	0.956	10.7
Post-transition	0.2	0.0873	0.994	7.9
	Musculoskeletal system			
Malthusian	0.95	0.0550	0.894	12.6
Early transition	0.24	0.0649	0.980	10.7
Late transition	0.07	0.0734	0.720	9.4
Post-transition	0.03	0.0828	0.925	8.4
	Neoplasms*			
Malthusian	1.1	0.0990	0.968	7.0
Early transition	1.5	0.0940	0.977	7.4
Late transition	1.4	0.0970	0.990	7.1
Post-transition	0.9	0.1022	0.998	6.8

Note: The parameter estimates are for Eq. 3.2; $\widehat{\alpha}$ is the anti-log of the estimate of A; doubling time is natural log of 2, or 0.693, divided by $\widehat{\beta}$

find out if the profiles of each category, as a degenerative one, show similar cross-regime change—and spot exceptions. Moreover, the cross-checks among the four sets help inspect if the broader pattern endures even as a different sets of people are examined.

Table 3.4 shows the average of the results across the sets, and Tables 3.5, 3.6, 3.7, and 3.8 have the results for each set I, II, III and IV. The fifth columns in all of them convey that the exponential curves fit the paths of most profiles well. Roughly 85% have R-square more than 90%, of which nearly all are more than 95%, reaching as high as 99% for many.

Table 3.6 Age-profile parameters of the sub-categories of non-communicable diseases, set II

Regime	$\widehat{\alpha}$	$\widehat{\beta}$	R^2	Doubling time (years)
	Circulatory system			
Malthusian	7.8	0.0790	0.994	8.8
Early transition	2.7	0.0942	0.980	7.4
Late transition	0.9	0.1060	0.986	6.5
Post-transition	0.4	0.1132	0.999	6.1
	Nervous system			
Malthusian	9.6	0.0343	0.950	20.2
Early transition*	7.4	0.0223	0.947	31.1
Late transition	0.7	0.0568	0.961	12.2
Post-transition	0.5	0.0624	0.984	11.1
	Digestive system			
Malthusian	10.4	0.0421	0.963	16.5
Early transition	5.7	0.0448	0.984	15.5
Late transition	0.4	0.0781	0.989	8.9
Post-transition	0.2	0.0906	0.967	7.7
	Musculoskeletal system			
Malthusian	0.8	0.0577	0.819	12.0
Early transition	0.2	0.0653	0.982	10.6
Late transition	0.02	0.0917	0.967	7.6
Post-transition	0.02	0.0887	0.969	7.8
	Neoplasms*			
Malthusian	1.3	0.0970	0.967	7.1
Early transition	1.7	0.0925	0.982	7.5
Late transition	1.3	0.0975	0.989	7.1
Post-transition	1.0	0.0985	0.999	7.0

Note: The parameter estimates are for Eq. 3.2; $\widehat{\alpha}$ is the anti-log of the estimate of A; doubling time is natural log of 2, or 0.693, divided by $\widehat{\beta}$

Marked by asterisks in Table 3.4 are the two main exceptions; and they appear as exceptions in all four sets (Tables 3.5, 3.6, 3.7 and 3.8). The fit is good for the early transition profile of the nervous system diseases, but its slope became flatter as the intercept fell. It pertains to that special case depicted in Fig. 3.3b, where the slope by itself can mislead (as does the resulting doubling time). They instead indicate a sizable gain in the aging potential. The proximate causes of the disorders of the nervous system—vital for human learning ability—may have reduced sizably for cohorts born in the 1870s and 1880s. The late and the post-transition profiles subsequently restored the general pattern of falling initial states and steeper slopes.

Table 3.7 Age-profile parameters of the sub-categories of non-communicable diseases, set III

Regime	$\widehat{\alpha}$	$\widehat{\beta}$	R^2	Doubling time (years)
	Circulatory system			
Malthusian	5.9	0.0830	0.992	8.4
Early transition	2.8	0.0931	0.981	7.4
Late transition	1.2	0.1038	0.994	6.7
Post-transition	0.4	0.1161	0.999	6.0
	Nervous system			
Malthusian	7.9	0.0345	0.974	20.1
Early transition*	7.3	0.0240	0.967	28.8
Late transition	1.2	0.0483	0.857	14.3
Post-transition	0.4	0.0624	0.982	11.1
	Digestive system			
Malthusian	4.8	0.0585	0.999	11.9
Early transition	4.2	0.0519	0.997	13.4
Late transition	2.7	0.0288	0.701	24.1
Post-transition	0.08	0.0752	0.805	9.2
	Musculoskeletal system			
Malthusian	0.6	0.0614	0.853	11.3
Early transition	0.2	0.0664	0.967	10.4
Late transition	0.05	0.0810	0.745	8.6
Post-transition	0.04	0.0816	0.923	8.5
	Neoplasms*			
Malthusian	1.5	0.0956	0.972	7.2
Early transition	1.6	0.0929	0.978	7.5
Late transition	1.5	0.0961	0.992	7.2
Post-transition	0.9	0.1013	0.998	6.8

Note: The parameter estimates are for Eq. 3.2; $\widehat{\alpha}$ is the anti-log of the estimate of A; doubling time is natural log of 2, or 0.693, divided by $\widehat{\beta}$

Meanwhile, by and large, the $\widehat{\alpha}_i$ in Table 3.4 fell across the regimes. The exception is the early transition profile for neoplasms, though the increase in $\widehat{\alpha}_i$ is small. Excepting neoplasms for which the profiles show minimal change, the growing $\widehat{\beta}_i$ indicate that the slopes in nearly all cases became steeper.

The change in aging potentials, however, varied across the categories. Between the Malthusian and the post-transition profiles, doubling times fell the most for the diseases of the digestive system (50%), followed by the nervous system (40%), circulatory system (34%), and the musculoskeletal system (33%). But change was negligible for neoplasms, and their doubling times have been relatively stable at

Table 3.8 Age-profile parameters of the sub-categories of non-communicable diseases, set IV

Regime	$\widehat{\alpha}$	$\widehat{\beta}$	R^2	Doubling time (years)
	Circulatory system			
Malthusian	7.9	0.0779	0.995	8.9
Early transition	2.4	0.0964	0.981	7.2
Late transition	1.0	0.1040	0.986	6.7
Post-transition	0.4	0.1118	0.999	6.2
	Nervous system			
Malthusian	9.6	0.0329	0.937	21.1
Early transition*	4.8	0.0296	0.960	23.4
Late transition	0.8	0.0536	0.919	12.9
Post-transition	0.5	0.0595	0.986	11.7
	Digestive system			
Malthusian	9.6	0.0430	0.960	16.1
Early transition	5.1	0.0470	0.988	14.7
Late transition	0.3	0.0789	0.987	8.8
Post-transition	0.2	0.0922	0.964	7.5
	Musculoskeletal system			
Malthusian	0.7	0.0574	0.808	12.1
Early transition	0.3	0.0585	0.981	11.9
Late transition	0.01	0.1056	0.997	6.6
Post-transition	0.03	0.0830	0.943	8.4
	Neoplasms*			
Malthusian	1.2	0.0983	0.965	7.1
Early transition	1.6	0.0929	0.982	7.5
Late transition	1.3	0.0971	0.989	7.1
Post-transition	0.9	0.0992	0.998	7.0

Note: The parameter estimates are for Eq. 3.2; $\widehat{\alpha}$ is the anti-log of the estimate of A; doubling time is natural log of 2, or 0.693, divided by $\widehat{\beta}$

about 7 years. All told, the cross-regime pattern in the sub-categories seems to echo that of the aggregate. The next section explores successive profiles since the nineteenth century.

3.5 So, Too, the Childhood Growth the Profiles Had Trailed

Tables 3.9, 3.10, 3.11, and 3.12 show the results for successive age-profiles for set I, II, III and IV since the early nineteenth century. Each set confirms four main features:

First, the fifth columns show that the parameters help track the age-profiles well; the fit for the vast majority of the cases is 99% or better. Second, $\widehat{\beta}_i$ were larger

Table 3.9 Parameters of successive cohort age-profiles in set I

Regime	Decade	$\widehat{\alpha}$	$\widehat{\beta}$	R^2	Doubling time (years)
Malthusian	1800–1809	15.6	0.0795	0.996	8.7
	1810–1819	15.2	0.0810	0.999	8.6
	1820–1829	17.7	0.0795	0.998	8.7
	1830–1839	23.2	0.0757	0.997	9.2
	1840–1849	29.1	0.0718	0.996	9.7
	1850–1859	33.4	0.0684	0.999	10.1
Early transition	1860–1869	25.6	0.0720	0.993	9.6
	1870–1879	21.2	0.0750	0.988	9.2
	1880–1889	17.3	0.0779	0.974	8.9
	1890–1899	12.4	0.0794	0.998	8.7
	1900–1909	12.4	0.0788	0.991	8.8
	1910–1919	11.3	0.0791	0.975	8.8
Late transition	1920–1929	7.1	0.0854	0.996	8.1
	1930–1939	6.3	0.0850	0.997	8.2
Post-transition	1940–1949	4.1	0.0906	0.992	7.7

Note: The parameter estimates are for Eq. 3.2; $\widehat{\alpha}$ is the anti-log of the estimate of A; doubling time is natural log of 2, or 0.693, divided by $\widehat{\beta}$

Table 3.10 Parameters of successive cohort age-profiles in set II

Regime	Decade	$\widehat{\alpha}$	$\widehat{\beta}$	R^2	Doubling time (years)
Malthusian	1795–1804	20.1	0.0736	0.987	9.4
	1805–1814	17.4	0.0780	0.989	8.9
	1815–1824	18.1	0.0785	0.998	8.8
	1825–1834	23.9	0.0747	0.999	9.3
	1835–1844	26.7	0.0731	0.994	9.5
	1845–1854	38.6	0.0661	0.998	10.5
	1855–1864	28.5	0.0705	0.990	9.8
Early transition	1865–1874	20.3	0.0751	0.989	9.2
	1875–1884	16.9	0.0766	0.997	9.0
	1885–1894	20.5	0.0714	0.979	9.7
	1895–1904	12.1	0.0793	0.987	8.7
	1905–1914	9.3	0.0826	0.983	8.4
	1915–1924	6.0	0.0888	0.997	7.8
Late transition	1925–1934	7.1	0.0834	0.996	8.3
	1935–1944	6.1	0.0831	0.998	8.3
Post-transition	1945–1954	4.9	0.0858	0.993	8.1

Note: The parameter estimates are for Eq. 3.2; $\widehat{\alpha}$ is the anti-log of the estimate of A; doubling time is natural log of 2, or 0.693, divided by $\widehat{\beta}$

Table 3.11 Parameters of successive cohort age-profiles in set III

Regime	Decade	$\widehat{\alpha}$	$\widehat{\beta}$	R^2	Doubling time (years)
Malthusian	1801–1810	20.1	0.0758	0.990	9.1
	1811–1820	19.6	0.0776	0.998	8.9
	1821–1830	21.4	0.0771	0.999	9.0
	1831–1840	28.7	0.0724	0.992	9.6
	1841–1850	39.3	0.0670	0.995	10.4
	1851–1860	34.0	0.0683	0.998	10.1
Early transition	1861–1870	26.9	0.0711	0.992	9.8
	1871–1880	23.0	0.0717	0.993	9.7
	1881–1890	19.5	0.0730	0.979	9.5
	1891–1900	12.6	0.0791	0.995	8.8
	1901–1910	11.6	0.0795	0.988	8.7
	1911–1920	9.4	0.0823	0.983	8.4
Late transition	1921–1930	6.9	0.0855	0.997	8.1
	1931–1940	6.1	0.0851	0.997	8.1
Post-transition	1941–1950	4.3	0.0895	0.994	7.7

Note: The parameter estimates are for Eq. 3.2; $\widehat{\alpha}$ is the anti-log of the estimate of A; doubling time is natural log of 2, or 0.693, divided by $\widehat{\beta}$

Table 3.12 Parameters of successive cohort age-profiles in set IV

Regime	Decade	$\widehat{\alpha}$	$\widehat{\beta}$	R^2	Doubling time (years)
Malthusian	1796–1805	18.2	0.0758	0.992	9.1
	1806–1815	17.0	0.0784	0.996	8.8
	1816–1825	16.8	0.0802	0.998	8.6
	1826–1835	21.4	0.0767	0.999	9.0
	1836–1845	28.3	0.0721	0.997	9.6
	1846–1855	39.1	0.0654	0.999	10.6
	1856–1865	30.6	0.0687	0.989	10.1
Early transition	1866–1875	19.6	0.0755	0.988	9.2
	1876–1885	15.5	0.0783	0.996	8.9
	1886–1895	17.2	0.0750	0.991	9.2
	1896–1905	15.7	0.0753	0.992	9.2
	1906–1915	14.2	0.0754	0.954	9.2
	1916–1925	6.9	0.0863	0.996	8.0
Late transition	1926–1935	7.2	0.0825	0.996	8.4
	1936–1945	6.2	0.0823	0.997	8.4
Post-transition	1946–1955	4.8	0.0855	0.996	8.1

Note: The parameter estimates are for Eq. 3.2; $\widehat{\alpha}$ is the anti-log of the estimate of A; doubling time is natural log of 2, or 0.693, divided by $\widehat{\beta}$

for the first two profiles of the nineteenth century, but then fell for about three decades through the 1850s. The $\widehat{\beta}_i$ then gradually improved, and the early transition profiles had more or less recovered to the level seen in early nineteenth century. The transition stage $\widehat{\beta}_i$ then trended up, though the change slowed or was setback for cohorts brought up in the inter-war years. Third, as the $\widehat{\beta}_i$ fell since the 1820s through the 1850s, their doubling times grew, and aging potentials worsened. Averaging about 8.8 years for the 1801s profiles, the doubling time worsened to about 10–10.5 years for the 1840s and 1850s profiles. They have shrunk since in nearly all profiles though there are periods of slow or no progress. Lastly, the series of $\widehat{\alpha}_i$ reaffirm the cycle for the Malthusian and the early-transition stage profiles, and the long-term progress that followed: the $\widehat{\alpha}_i$ for cohorts from the first two decades of the nineteenth century were relatively small, then worsened to peak for the 1840s and 1850s profiles (high α is bad); the $\widehat{\alpha}_i$ of the 1880s and 1890s profiles were similar to the ones from the profiles from the early nineteenth century.

If each cohort were a generation, then the first half of the nineteenth century is that rare episode when a series of three to four generations fared worse than the preceding ones. People brought up since about the 1820s through the 1850s had an increasingly poorer start and may have suffered severe setbacks as children.

To explore the matter further, Fig. 3.4 juxtaposes the long-term trends in average adult stature, and in the two parameters of the age-profiles for set I. In periods when the average height of a cohort was poorer than that of the preceding cohort, the age-profiles began in worse shape; stayed more elevated over the next 50 years; became flatter; and displayed lesser aging potentials. As the trend in children's growth reversed after the mid-nineteenth century, the age-profiles improved as well.

The profile parameters and average heights are two very different sources of evidence. Each one was also measured at different times. The parameters can be calculated only after each cohort had reached 74 years of age, about five decades after the average height was attained. They both, however, indicate strongly toward linkages with childhood development.

To reconnect with the debate on aging, as the new trends in older people appeared in the 1960s, it gradually became clearer that spearheading the trends were a newer set of cohorts, whose childhood physiological development may have been better than the preceding cohorts. Anthropometric historians uncovered heights children were able to attain over the past two centuries. Comparing distant cohorts, they found that measures of children's development such as height and body mass index were related to lesser incidence, fewer deaths, and better survival rates at middle-to-older ages (e.g. Waaler 1984; Fogel 1993; Steckel and Costa 1997; Costa 2002, 2003; Floud et al. 2011). On that basis, scholars surmised that the new time trends in the middle-to-older age people could continue if newer generations with better childhood development enter the picture. The experience for mid-nineteenth century cohorts in Figs. 3.4a, b suggests that the situation can be thrown into reverse as well. Chapters 8 and 9 will revisit this issue along with the life-course factors in aging. Chapter 10 will then study both childhood-linked and

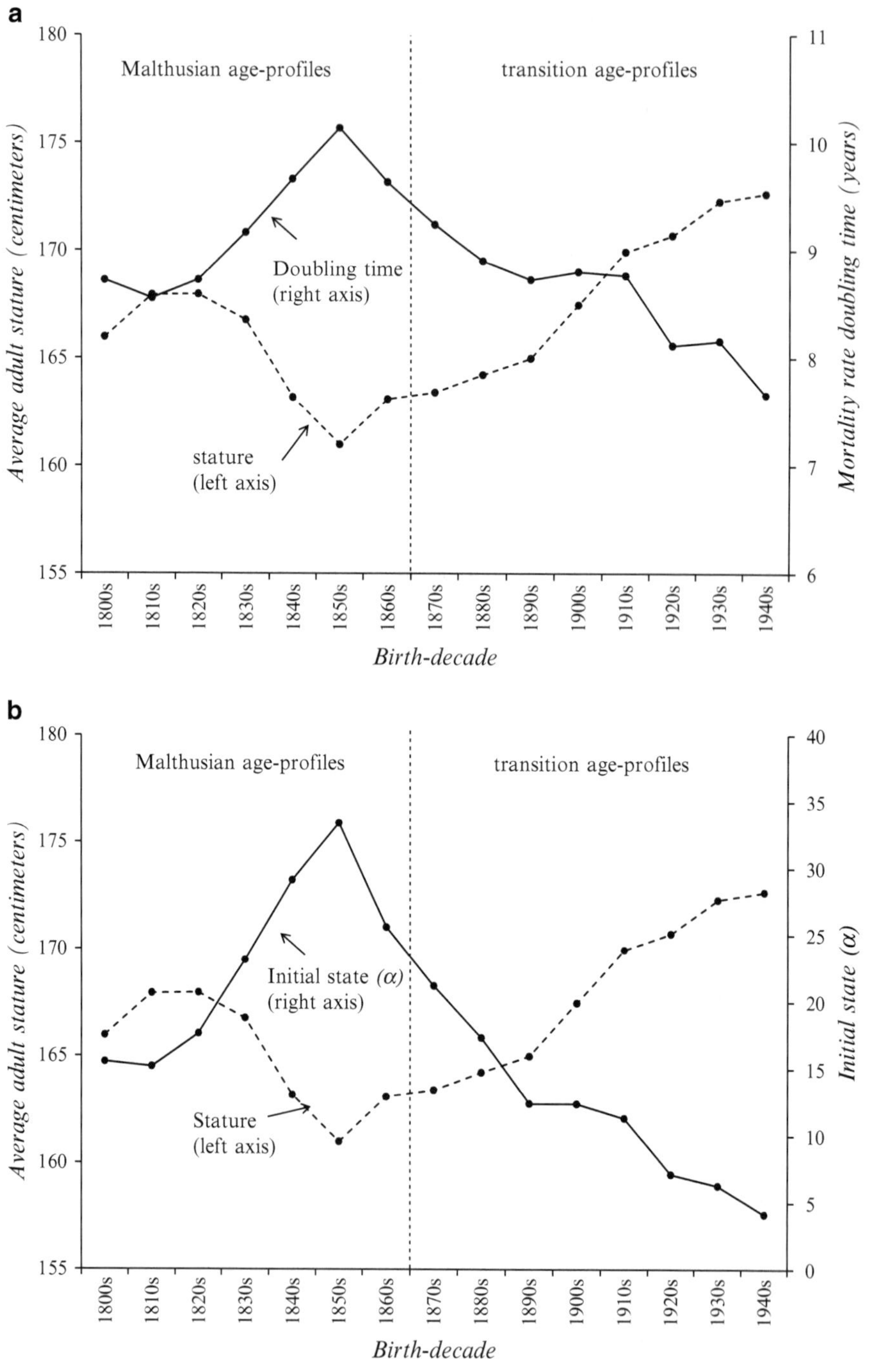

Fig. 3.4 The long-term trends in average adult stature of cohorts and their profile parameters (**a**) Doubling times (**b**) Initial state (Sources: HMSO; Floud et al. (1990), Rosenbaum (1988); author's calculations)

life-course factors in analytical models of the per capita healthcare spending profiles.

Further, the downturn in stature in the first half of the nineteenth century has been at the center of the long-running debate on living standards during industrialization (Hobsbawm 1957; Hartwell and Engerman 1975; Floud and Wachter 1982; Von Tunzelmann 1979, 1985; Lindert and Williamson 1983; Lindert 1994; Komlos 1993a, b, c, 1998b; Steckel 1995; Steckel and Floud 1997), and of the analytical frameworks for measuring the standard (Sen 1984, 1985, 1987, 1999).

"The standard of life cannot be so detached from the nature of the life the person leads" (Sen 1987, p. 8). Quoting Sen in the same reference, Geoffrey Hawthorn points out that "the value of the living standard lies in the living." The emphasis on 'living' has led scholars, most notably A. K. Sen, to explain this complex, often ineffable, issue in terms of 'features of the state of existence of a person,' or of 'functioning bundles' and 'capabilities,' including the 'capabilities to choose,' and not just commodities a person or household can own or produce. Height, for example, reflects the 'state of existence' of children. It reflects how commodities and the distribution of commodities influence children's development.

Likewise, aging potentials summarize the age-profiles, expressing their 'state of existence' or their 'capabilities' in terms of the rates of non-communicable diseases over 50 years. They express the extent of loss in the capabilities to function or in the 'functioning bundles.' For instance, the profiles of successive cohorts brought up during 1820–1850 had deteriorated. Some of those profiles had lasted through the first quarter of the twentieth century, well beyond the time when the usual measures of living standards such as literacy rates, life expectancies, per capita real incomes, and even stature, had been indicating that the living standard may have improved.

So, even as some measures of the standard may have improved, the 'functioning capabilities' may have deteriorated. That brings up a puzzle: *when* should one say that the standard of living was worsening or improving? Other dimensions of the standard are more clear-cut about the issue than the profiles as each profile lasted 50 years. For instance, period-life expectancies, indicate that wellbeing improved since the last quarter of the nineteenth century. Trends in stature indicate that the standard likely improved since the 1860s. However, those inferences about the standard remain muddy as the rates of non-communicable diseases were climbing in the second half of the nineteenth century, and the cohort age-profiles had been shifting upward.

This difference in timing across the measures suggest that various aspects of the standard should not be expected to move in tandem. There can be times when more of them indicate improvement than at other times. As will become clearer in Chap. 5, the age-profiles of infectious diseases were downward-sloping. The aspect of the living standard related to infectious diseases had been improving since at least the last quarter of the nineteenth century. However, that coincided for some time with worsening profiles of non-communicable diseases. Thus, the profiles across the two broad disease categories are likely to yield a mixed picture as well.

Furthermore, some aspects of capabilities relate in a clear-cut way to choice, while others do not (Williams 1987). Many non-communicable diseases pertain to involuntary organs. Even if one presumes that people know perfectly why, and by how much, the organs may be deteriorating (a very strong presumption), the organs cannot necessarily be willed into perfect functioning, at least not until an effective medical therapy is at hand. The upward slope of the age-profiles of most non-communicable diseases, themselves, has a sense of involuntariness about it: senescence becomes more potent with age; death rates increase with age. This feature could be seen as what Sen (1987) calls an 'absolute notion in the space of capabilities.' The limiting absolute profile would be the ideal case, where rates are negligible until the maximal age and the slope is the steepest possible.

However, there is a relative dimension that allows cross-cohort comparison. While all age-profiles slope upward in the 'absolute' sense, some are steeper than others. Based on their initial states and slopes, one may infer about functional capabilities of one cohort versus another, or about how far one cohort was from the ideal than another. The sub-categories of non-communicable diseases could allow even more granular comparisons.

The third main issue in the concept of 'capabilities' is of the 'sets of co-realizable capabilities' (Nussbaum 2011). It covers socio-political conditions in which people may realize several types of capabilities, or at least acquire the means to realize them. Chapters 6 and 7 will discuss the socio-economic and political contexts in which the age-profiles had shifted. Children brought up under extreme political inequality, for instance, had some of the worst age-profiles. The lack of franchise (and of knowledge) may have left the mass of households less capable of improving the conditions in which their children were brought up.

References

Costa, D. L. (2002). Changing chronic disease rates and the long-term declines in functional limitations among older men. *Demography, 39*(1), 119–137.

Costa, D. L. (2003). Understanding the mid-life and older-age mortality declines: Evidence from the Union Army veterans. *Journal of Econometrics, 112*(1), 175–192.

Finch, C. E. (1990). *Longevity, senescence and the genome*. Chicago: The University of Chicago Press.

Floud, R., & Wachter, K. (1982). Poverty and physical stature: Evidence on the standard of living in London boys, 1770–1870. *Social Science History, 6*, 422–452.

Floud, R., Wachter, K., & Gregory, A. (1990). *Height, health, and history: Nutritional status in the United Kingdom, 1750–1980*. Cambridge: Cambridge University Press.

Floud, R., Fogel, R. W., Harris, B., & Hong, S. C. (2011). *The changing body, health nutrition and human development in the western world since 1700* (NBER). New York: Cambridge University Press.

Fogel, R. W. (1993). Economic growth, population theory and physiology: The bearing of long-term processes on the making of economic policy. *American Economic Review, 84*(3), 369–395.

Fridlizius, G. (1989). The deformation of cohorts: Nineteenth century mortality decline in generational perspective. *Scandinavian Economic History Review/Economy and Society, 37*, 3–17.

Fries, J. F. (1980). Aging, natural death and compression of morbidity. *New England Journal of Medicine, 303*(3), 130–135.
Fries, J. F. (1989). The compression of morbidity: Near or far? *Milbank Memorial Fund Quarterly, 67*(2), 208–232.
Gavrilov, L. A., & Gavrilova, N. S. (2011). Mortality measurement at advanced ages: A study of the social security administration death master file. *North American Actuarial Journal, 15*(3), 432–447.
Gavrilova, N. S., Gavrilov, L. A., Evdokushkina, G. N., & Semyonova, V. G. (2003). Early-life predictors of human longevity: Analysis of the 19th century birth-cohorts. *Annales de Demographie Historique, 2*, 177–198.
Gompertz, B. (1825). On the nature of the function expressive of the law of human mortality, and on the mode of determining the value of life contingencies. *Philosophical Transactions of the Royal Society, 115*, 513–585.
Gompertz, B. (1872). On the uniform law of mortality from birth to extreme old age, and on the law of sickness. *Journal of the Institute of Actuaries, 16*, 329–344.
Harris, B. (2001). "The child is the father to man" the relationship between child health and adult mortality in the 19th and 20th centuries. *International Journal of Epidemiology, 30*, 688–696.
Hartwell, R. M., & Engerman, S. (1975). Models of immiserization: The theoretical basis of pessimism. In A. J. Taylor (Ed.), *The standard of living in Britain in the industrial revolution*. London: Methuen.
Hayflick, L. (1985). Theories of aging. In R. Andres, E. L. Bierman, & W. R. Hazzard (Eds.), *Principles of geriatric medicine* (pp. 9–21). New York: McGraw Hill.
Hobsbawm, E. J. (1957). The British standard of living, 1790–1850. *The Economic History Review, 10*, 46–68.
Horiuchi, S. (1989). Some methodological issues in the assessment of the deceleration of the mortality decline. In L. Ruzicka, G. Wunsch, & P. Kane (Eds.), *Differential mortality: Methodological issues and biosocial factors* (pp. 64–78). Oxford: Clarendon Press.
Kermack, W. O., McKendrick, A. G., & McKinlay, P. L. (1934, March). Death rates in Great Britain and Sweden; Some general regularities and their significance. *The Lancet, 31*, 698–703.
Kirkwood, T. B. L. (1977). Evolution of ageing. *Nature, 270*, 301–304.
Kirkwood, T. B. L. (1981). Repair and its evolution: Survival versus reproduction. In C. R. Townsend & P. Calow (Eds.), *Physiological ecology an evolutionary approach to resource use* (pp. 165–189). Oxford: Blackwell Scientific Publications.
Kirkwood, T. B. L. (1985). Comparative evolutionary aspects of longevity. In C. E. Finch & E. L. Schneider (Eds.), *Handbook of the biology of aging* (2nd ed., pp. 27–45). New York: Plenum Press.
Kirkwood, T. B. L. (1999). *Time of our lives: The science of human aging*. Oxford: Oxford University Press.
Kirkwood, T. B. L., & Holliday, R. (1979). The evolution of longevity. *Proceedings of the Royal Society B, 205*, 532–546.
Komlos, J. (1993a). Further thoughts on the nutritional status of the British population. *The Economic History Review, 46*(3), 363–366.
Komlos, J. (1993b). A Malthusian episode revisited: The height of British and Irish servants in colonial America. *The Economic History Review, 46*(2), 768–782.
Komlos, J. (1993c). The secular trend in the biological standard of living in the United Kingdom: 1730–1860. *The Economic History Review, 46*(1), 115–144.
Komlos, J. (1998). Shrinking in a growing economy: The mystery of physical stature during the industrial revolution. *Journal of Economic History, 58*(3), 779–802.
Lee, R. D., & Carter, L. R. (1992). Modeling and forecasting United States mortality. *Journal of the American Statistical Association, 87*(419), 659–671.
Lindert, P. (1994). Unequal living standards. In R. Floud & D. McCloskey (Eds.), *The economic history of Britain since 1700* (Vol. 1, pp. 357–386). Cambridge: Cambridge University Press.

Lindert, P., & Williamson, J. (1983). English workers' living standards during the industrial revolution: A new look. *The Economic History Review, 36*, 1–25.

Medawar, P. B. (1952). *An unsolved problem of biology*. London: Lewis.

Nussbaum, M. (2011). *Creating capabilities, the human development approach*. Cambridge, MA: The Belknap Press of Harvard University Press.

OECD. (2013). *Public spending on health and long-term care: A new set of projections* (OECD economic policy papers, no. 6). Paris: OECD.

Oeppen, J., & Vaupel, J. (2002). Broken limits to life expectancy. *Science, 296*, 1029–1031.

Olshansky, S. J. (1985). Delay versus elimination of degenerative diseases. *American Journal of Public Health, 75*(7), 754–757.

Olshansky, S. J., & Ault, A. B. (1986). The fourth stage of the epidemiologic transition: The age of delayed degenerative diseases. *Milbank Memorial Fund Quarterly, 64*(3), 355–391.

Olshansky, S. J., Carnes, B. A., & Cassel, C. (1990). In search of Methuselah: Estimating the upper limits to human longevity. *Science, 250*, 634–640.

Rau, R., Soroko, E., Jasilionis, D., & Vaupel, J. (2008). Continued reductions in mortality at advanced ages. *Population and Development Review, 34*(4), 747–768.

Rose, M. R. (1991). *Evolutionary biology of aging*. New York: Oxford University Press.

Rosenbaum, S. (1988). 100 years of heights and weights. *Journal of the Royal Statistical Society, 151*, 276–309.

Sen, A. K. (1984). *The living standard. Oxford Economic Papers*, 36.

Sen, A. K. (1985). *Commodities and capabilities*. Amsterdam: North-Holland.

Sen, A. K. (1987). In G. Hawthorn (Ed.), *The standard of living*. Cambridge: Cambridge University Press.

Sen, A. K. (1999). *Commodities and capabilities*. Oxford: Oxford University Press.

Spicer, C. C. (1954). The generation method of analysis applied to mortality from respiratory tuberculosis. *Journal of Hygiene, 52*(3), 361–368.

Springett, V. H. (1952). An interpretation of the statistical trends in tuberculosis. *Lancet, 259*(6708), 521–525.

Steckel, R. H. (1995). Stature and the standard of living. *Journal of Economic Literature, 33*(4), 1903–1940.

Steckel, R. H., & Costa, D. L. (1997). Long-term trends in health, welfare and economic growth in the United States. In R. H. Steckel & R. Floud (Eds.), *Health, welfare and industrialization*. Chicago: University of Chicago Press.

Steckel, R. H., & Floud, R. (1997). *Health, welfare and industrialization*. Chicago: University of Chicago Press.

Strehler, B. L., & Mildvan, A. S. (1960). General theory of mortality and aging. *Science, 132*(3418), 14–21.

Thatcher, A. R., Kannisto, V., & Vaupel, J. W. (1998). *The trajectory of mortality from ages 80 to 120*. Odense: Odense University Press.

Vaupel, J. W., & Lundström, H. (1994). Prospects for longer life expectancy. In D. Wise (Ed.), *Studies in economics of aging*. Chicago: University of Chicago Press.

Von Tunzelmann, N. (1979). Trends in real wages, 1750–1850, revisited. *The Economic History Review, 32*, 33–49.

Von Tunzelmann, N. (1985). The standard of living debate and optimal economic growth. In J. Mokyr (Ed.), *The economics of the industrial revolution*. London: Allen and Unwin.

Waaler, H. T. (1984). Height, weight and mortality: The Norwegian experience. *Acta Medica Scandinavia Supplement, 679*, 1–51.

Williams, G. C. (1957). Pleiotropy, natural selection and the evolution of senescence. *Evolution, 11*(4), 398–411.

Williams, B. (1987). The standard of living: Interests and capabilities. In G. Hawthorn (Ed.), *Amartya Sen's The standard of living*. Cambridge: Cambridge University Press.

Chapter 4
Resolving the Paradox of Aging

4.1 Requires Linking Cohort-Profiles to the Aggregate

Since the 1820s or so, cohort-profiles had first shifted up (for the worse) before shifting down (for the better) over the long-term. Meanwhile, the long-term path of the aggregate rate in Fig. 4.1 shows that it grew over the second half of the nineteenth century before receding over the long term.

How were the shifting profiles linked to the aggregate's path? As Chap. 1 pointed out, contributions by many cohorts make up a period-profile, which then sums up to the aggregate rate. For instance, the age groups and the cohorts that contributed to the aggregate rate in 1894 were: 25–34 from the 1860s profile, 35–44 from the 1850s, 45–54 from the 1840s, 55–64 from the 1830s, and 65–74 from the 1820s profile. They were mostly Malthusian regime cohorts, each at a different stage of its life cycle. A decade later in 1904, the 25–34-year olds were from the 1870s early-transition stage profile, while people of age 35–44, 45–54, 55–64 and 65–74 were from the 1860s, 1850s, 1840s and 1830s profiles.

To see how the aggregate's direction is linked to the way the profiles were shifting, their contributions to the aggregate have to be isolated. For each year on the horizontal axis of Fig. 4.2a, the segments of each bar show the contributions by the profiles from each regime. For year t the share of regime j's profiles is:

$$s_{jt} = \sum\nolimits_{i=1}^{n} w_{it}^{j}\, m_{it}^{j} \div \sum\nolimits_{j=1}^{r} \sum\nolimits_{i=1}^{n} w_{it}^{j}\, m_{it}^{j}\,, \tag{4.1}$$

where $j = 1,2,\ldots, r$ denote the regimes; and each regime can have $i = 1,2,\ldots, n$ profiles. The numerator is the deaths in profile i from regime j (m_{it}^{j}) multiplied by its weight in the population of 25–74-year olds (w_{it}^{j}). The denominator is the weighted sum of all profiles contributing in the year. The same calculation for set II is shown in Fig. 4.2c. To track the aggregate's direction along with the changing contributions in the bars, an index of the age-standardized rate (with the first one as 100) is noted on the top of each bar.

S. Arora, *The Transitions of Aging*, International Perspectives on Aging 12,
DOI 10.1007/978-3-319-14403-0_4

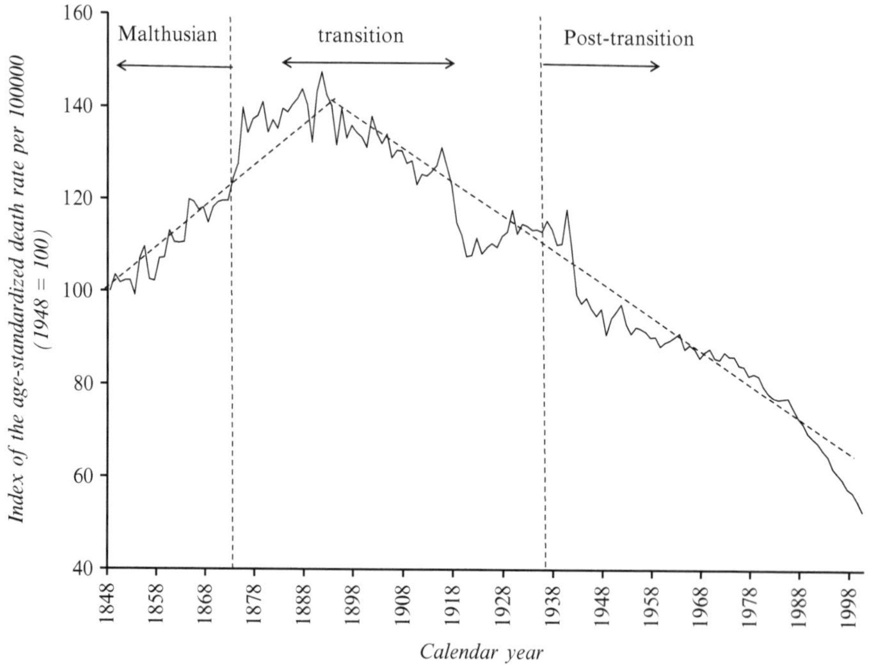

Fig. 4.1 Aggregate rate of non-communicable diseases, age 25–74, 1848–2000 (Sources: HMSO; author's calculations)

The aggregate rate would grow if the mix of profiles contributing to it worsens. The mix, in turn, worsens as the better profiles leave, and the ones entering the mix were worse or had shifted up. Likewise, it would fall if the poorer profiles leave, and the ones joining the mix were better or had shifted down.

The upward shifts in the series of profiles since the 1820s had worsened the mix contributing to the aggregate rate. As people in those profiles were aging, at each age they were posting heftier rates than the ones from the preceding profiles had posted. Figure 4.2b zooms into the period. It isolates the contributions of the profiles before the 1820s; of the 1820s–1850s profiles; of the 1860s profile; and of the early-transition 1870s profile. Between 1854 and 1884, as the mix evolved toward the 1820s–1850s profiles, and their share expanded from 18% to 84%, the aggregate rate grew by 30%. It then stayed elevated over the next two decades. The better 1860s and 1870s profiles were still too young and their contributions not yet large enough to sway the aggregate rate. Figure 4.2d reaffirms the pattern for set II.

As the mix gradually improved since, the long-term aggregate rate trended down. The Malthusian profiles were slowly leaving and the new transition-stage profiles began increasingly swaying the aggregate rate. It fell about 18% between 1894 and 1924 as the early-transition profiles' contributions grew from zero to 49% and those of the Malthusian ones fell. By 1944, the rate fell another 10% as the

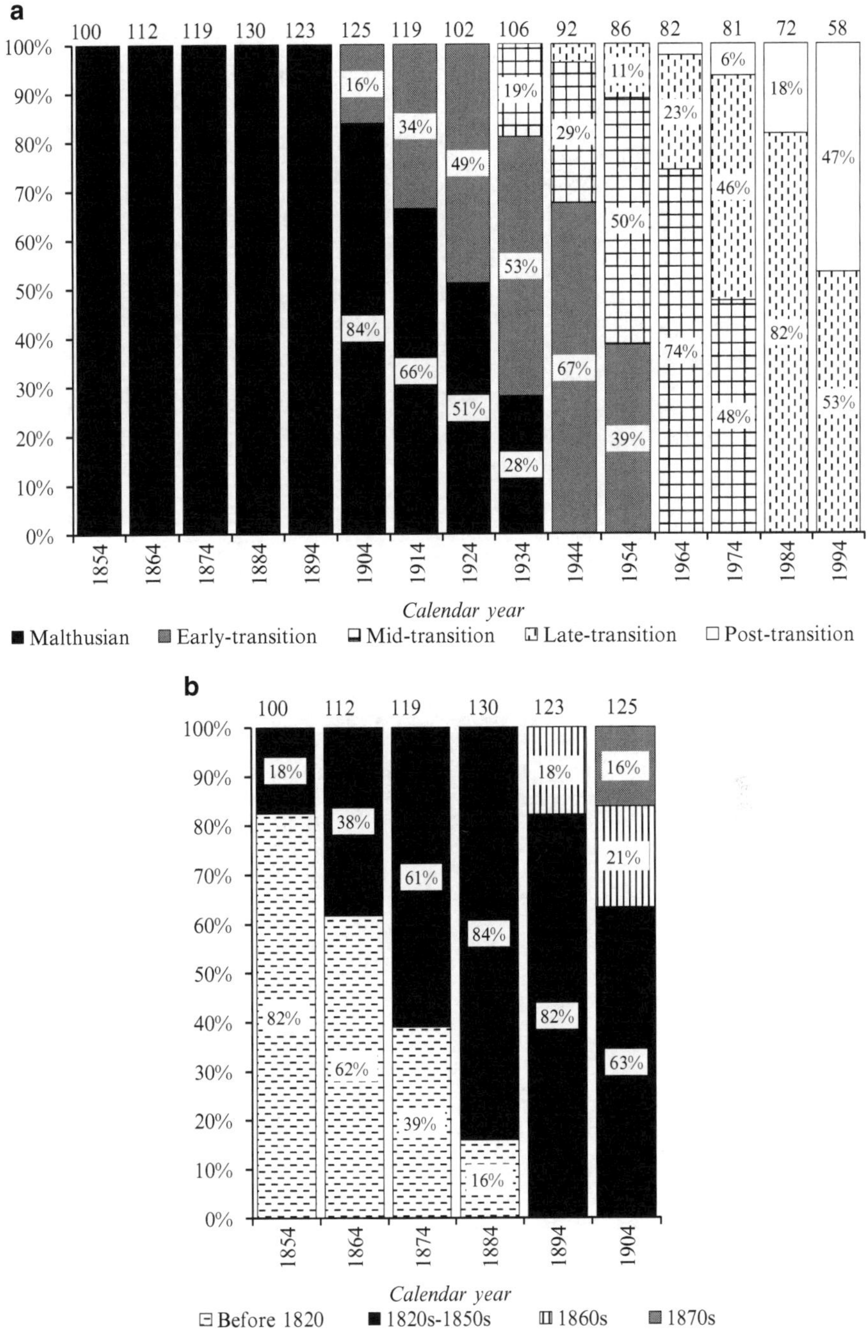

Fig. 4.2 The contributions of cohort-profiles to period-aggregate rates of non-communicable diseases (**a**) Set I (**b**) Set I Malthusian profiles (**c**) Set II (**d**) Set II Malthusian profiles (Sources: HMSO; author's calculations)

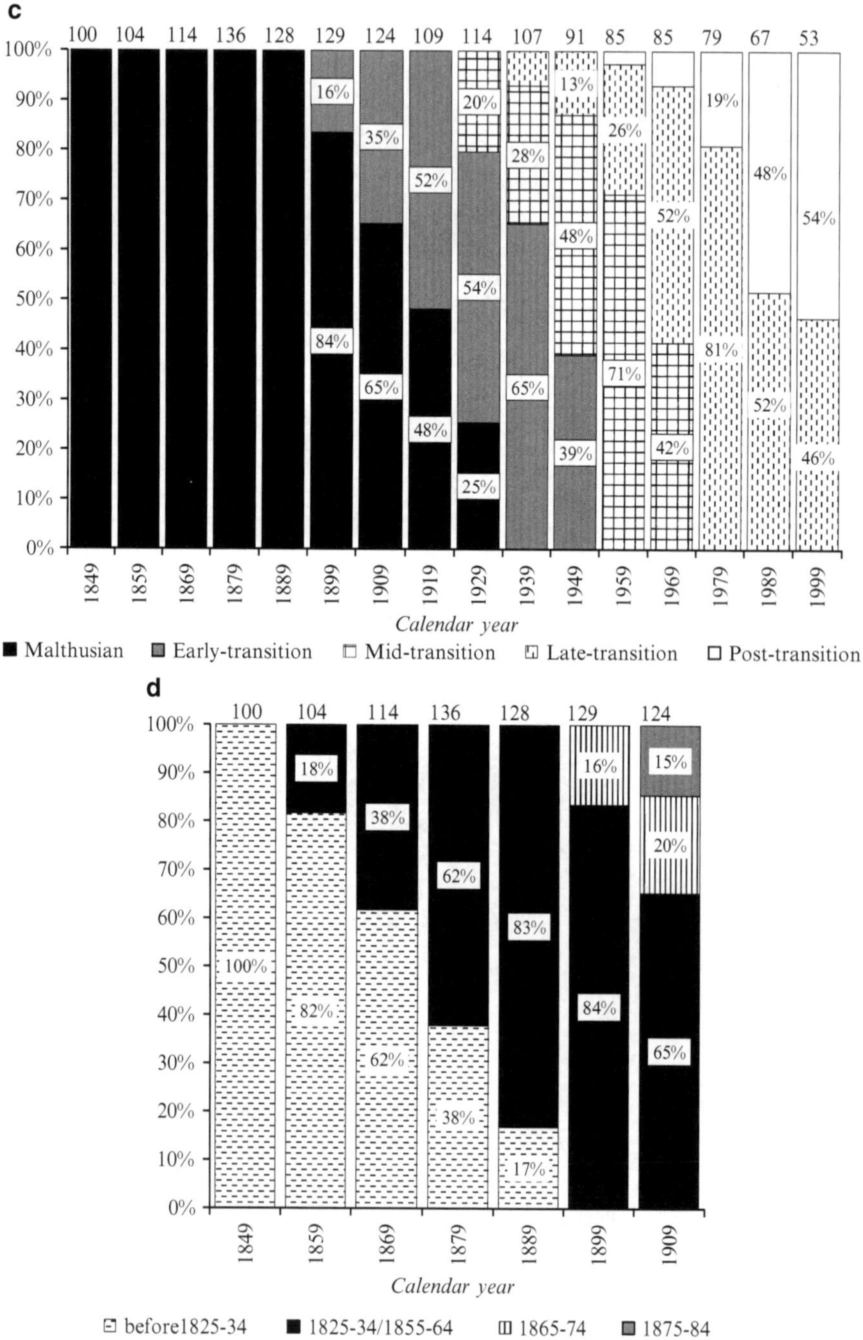

Fig. 4.2 (continued)

share of the early-transition profiles reached 67% and that of the mid-transition ones (1890s, 1901s) was 30%. By 1974, well into the 'age of delayed deaths from degenerative diseases,' the mix of profiles was mainly mid-transition (1890s, 1901s) and late-transition (1910s, 1920s).

4.2 And Simulating the Aggregate's Path

Apart from indicating the direction of the aggregate rate, can the cohort-profile parameters, α and β, help simulate the rate itself? This can be found out in four steps: (1) generate the rates each cohort's parameters imply for each age in their profiles; (2) line up in each calendar year the profiles contributing to the aggregate rate (i.e. make the period-profile for each year); (3) sum the implied death rates by age to form a synthetic or an implied aggregate rate; and (4) compare this synthetic or implied aggregate rate to the actual one for the year to find out how far off they were.

The exercise serves at least three ends. First, it tests how well the parameters can help track the actual historical series. One reason to care about the issue is parsimony: instead of several points along each profile, just two parameters may be able to describe the profiles' own paths as well as the path of the aggregate series they help form.

Second, it clarifies the links between cohort-profiles and period-profiles. How well the cohort parameters help track the aggregate rate hinges on how closely they can mimic the period age-specific rates and thus the period-profiles. This will play a key role later in isolating the childhood-linked and life-course related aspects of the period-profiles. At this stage, however, the main issue is whether the cohort parameters can help simulate the period-profiles.

This very matter is also at the core of the alternative outlooks of aging and the prospects of healthcare spending outlined in Chap. 1. They are mainly about the locations and shapes of the future period-profiles; and about the likely α and β of the future period-profiles. The brighter outlook of aging says that newer cohorts will fare better and the period-profiles they help form would be better than the current one; whereas, the grimmer outlook cautions that the newer cohorts may not have an edge over existing ones, and would form period-profile not so different from the current one; or α and β of the future profiles will be more or less remain the same (static profile). Likewise for the projections of the spending profiles: a static profile presumes that the period-profile of future healthcare spending is likely to stay unchanged, which means that the cohorts contributing to that spending profile are likely to be more or less identical. The simulation will thus help explore under what conditions the underpinnings of such scenarios may play out.

Third, the links between cohort- and period-profiles may help find out in what circumstances the aggregate rate of non-communicable diseases grows, shrinks, or stagnates over the long term, and will help shed some light on the paradox of aging.

Table 4.1 The implied versus the actual aggregate rate, 1874–1904

Cohort	Profile parameters			Age	Implied rate	Actual rate
	$\hat{\alpha}$	$\hat{\beta}$	*Age*		1874	1874
1840–1849	29.13	0.0718	29.5	25–34	241.9	229.2
1830–1839	23.16	0.0757	39.5	35–44	259.7	484.6
1820–1829	17.66	0.0795	49.5	45–54	901.6	904.8
1810–1819	15.20	0.0810	59.5	55–64	1,887.7	1,834.5
1800–1801	15.56	0.0795	69.5	65–74	3,899.5	4,208.7
					1,225.8	**1,265.1**
	$\hat{\alpha}$	$\hat{\beta}$	*Age*		1884	1884
1850–1859	33.40	0.0684	29.5	25–34	251.3	264.0
1840–1849	29.13	0.0718	39.5	35–44	495.7	554.5
1830–1839	23.16	0.0757	49.5	45–54	979.6	1,041.8
1820–1829	17.66	0.0795	59.5	55–64	1,995.7	2,022.8
1810–1819	15.19	0.0810	69.5	65–74	4,245.3	4,407.7
					1,320.3	**1,377.9**
	$\hat{\alpha}$	$\hat{\beta}$	*Age*		1894	1894
1860–1869	25.61	0.0720	29.5	25–34	214.0	227.3
1850–1859	33.40	0.0684	39.5	35–44	498.0	480.3
1840–1849	29.13	0.0718	49.5	45–54	1,015.9	967.4
1830–1839	23.16	0.0757	59.5	55–64	2,087.4	1,946.3
1820–1829	17.66	0.0795	69.5	65–74	4,417.5	4,292.6
					1,358.6	**1,307.0**
	$\hat{\alpha}$	$\hat{\beta}$	*Age*		1904	1904
1870–1879	21.20	0.0750	29.5	25–34	194.0	192.9
1860–1869	25.61	0.0720	39.5	35–44	439.6	428.0
1850–1859	33.40	0.0684	49.5	45–54	987.0	951.2
1840–1849	29.13	0.0718	59.5	55–64	2,082.1	2,048.6
1830–1839	23.16	0.0757	69.5	65–74	4,447.9	4,512.3
					1,337.8	**1,331.3**

Note: The implied rate is $\left(\hat{\alpha} \times e^{\hat{\beta}\ age}\right)$. The aggregate rate is age-adjusted using weights of the age-groups in the population of 25–74-year olds in 1994

Before presenting the results, it is best to illustrate how the simulation works by selecting two episodes germane to the outlooks. Using the α and β of the profiles from set I, Table 4.1 illustrates the links between the cohorts and the aggregate rate over the less-known episode over 1874–1904 when the aggregate rate was climbing. It sheds light on the grimmer scenario, the expansion of deaths. Table 4.2 studies the better-known episode over 1954–1984, when the shrinking aggregate rate had prompted the debate on aging. It exposes the brighter scenario of aging, the compression.

In the top quarter segment of Table 4.1, the last column labeled 'Actual rate' has the crude rates across the age groups as reported in 1874. They make up the period-

Table 4.2 The implied versus the actual aggregate rate, 1954–84

Cohort	Profile parameters			Age	Implied rate	Actual rate
	$\hat{\alpha}$	$\hat{\beta}$	*Age*		1954	1954
1920–1929	7.13	0.0854	29.5	25–34	88.5	87.5
1910–1919	11.30	0.0791	39.5	35–44	257.0	191.2
1900–1909	12.39	0.0788	49.5	45–54	611.8	530.0
1890–1899	12.41	0.0794	59.5	55–64	1,401.5	1,387.4
1880–1889	17.33	0.0779	69.5	65–74	3,892.5	3,564.4
					1,000.3	**918.1**
	$\hat{\alpha}$	$\hat{\beta}$	*Age*		1964	1964
1930–1939	6.27	0.0850	29.5	25–34	76.9	81.2
1920–1929	7.13	0.0854	39.5	35–44	208.0	193.0
1910–1919	11.30	0.0791	49.5	45–54	566.9	523.3
1900–1909	12.39	0.0788	59.5	55–64	1,345.1	1,361.6
1890–1899	12.41	0.0794	69.5	65–74	3,102.0	3,288.7
					852.9	**871.2**
	$\hat{\alpha}$	$\hat{\beta}$	*Age*		1974	1974
1940–1949	4.14	0.0906	29.5	25–34	59.9	69.3
1930–1939	6.27	0.0850	39.5	35–44	179.9	171.6
1920–1929	7.13	0.0854	49.5	45–54	488.6	541.8
1910–1919	11.30	0.0791	59.5	55–64	1,250.1	1,335.6
1900–1909	12.39	0.0788	69.5	65–74	2,957.3	3,237.7
					789.5	**855.6**
	$\hat{\alpha}$	$\hat{\beta}$	*Age*		1984	1984
1950–1959			29.5	25–34	65.3*	65.3
1940–1949	4.14	0.0906	39.5	35–44	148.1	134.8
1930–1939	6.27	0.0850	49.5	45–54	420.9	424.7
1920–1929	7.13	0.0854	59.5	55–64	1,148.0	1,236.3
1910–1919	11.30	0.0791	69.5	65–74	2,757.0	2,957.4
					723.9	**765.1**

Note: The implied rate is $\left(\hat{\alpha} \times e^{\hat{\beta}\ age}\right)$. The aggregate rate is age-adjusted using weights of the age-groups in the population of 25–74-year olds in 1994. The asterisk stands for actual rate

profile of the crude rates, and the rates sum up to the age-adjusted aggregate rate 1,265.1 per 100,000.

Meanwhile, the column labeled 'Implied rate' has the rates produced by the stated parameter values of the cohorts in the first column. Five cohorts contribute to each period-profile. For example, the rate for age 25–34 is from the 1840–1849 cohort because it was 25–34 years old in 1874. Each implied rate is the product $\hat{\alpha} \times e^{\hat{\beta} \times age}$, where e is the exponential term (about 2.71828). The mix of cohorts mentioned in the first column shaped the simulated period-profile or the simulated period-rates for 1874. Those period rates, in turn, sum to the simulated age-adjusted aggregate rate of 1,225.8 per 100,000, about 3.1% short of the Actual aggregate rate.

Repeat the exercise in 1884, but now with a different cohort mix. By this time the 1800–1809 cohort has left the mix as it does not contribute to the aggregate rate of the 25–74-year olds anymore. It was one of the better profiles from the early nineteenth century before the series of upward shifts had begun. Its exit worsens the cohort mix. Meanwhile, the 1850–1859 profile has now entered the picture as it was 25–34 years old in 1884. At age 25–34, it was in poorer shape than the 1840–1849 profile. Its entry worsened the mix of profiles: it began by posting 251.3, a rate more than 241.9 the 1840s profile had registered when it was 25–34 in 1874. Meanwhile, the 1840–1849 profile had now matured to age 35–44 in 1884 and posted a higher rate than the 35–44-year olds of the 1830s profile had posted in 1874. The heftier rates in 1884 than in 1874 are a result of the upward shifts in the cohort-profiles. They boosted the simulated aggregate to 1,320.3, which was about 4.2% off the actual 1,377.9. Built in a similar way, the climbing simulated aggregate rate was off the actual one by about 4% in 1894, and about 0.5% in 1904.

Table 4.1 clarifies that if the outlook for aging is grim and if it projects a growing aggregate rate over the long-term, then upward shifts in the period-profiles should be central to the outlook. That is, successive cohorts should fare increasingly worse. Over a period of five to six decades, it amounts to saying that the profile of your grandchildren's generation would fare worse than the profile of your children's generation, which, in turn will be worse than your own. Further, the profiles should be becoming increasingly flatter, or their βs should be shrinking.

Table 4.2 illustrates the exercise for the period after World War II when the aggregate rate had long been falling. In 1954, 1964 and 1974 the contributing profiles were mainly the ones that had shifted downward in the transition stage. Their lesser rates at each age were summing up to a falling aggregate. The simulated aggregate in 1984 is based on partial data; the 1950s profile would have been 25–34 years old then, but it had not aged enough for it parameters to be estimated. So the simulated rate in 1984 is based on the actual reported rate for that age group, marked by the asterisk, combined with the implied rates for the rest of the contributing cohorts.

This episode exposes the makings of a brighter outlook of aging. If the outlook expects the aggregate rate to fall, it must also expect the period-profiles to shift downward and become steeper. That could happen, in turn, if the successive cohorts shaping the period-profiles fare better. It exemplifies the compression of deaths from the vantage point of cohorts.

Based on the workings in Tables 4.1 and 4.2, one may conjecture that the period-profiles in any year could have been at least partially echoing the past. If the state of children tells on how they may fare as adults, a period-profile could signal the setbacks or gains in children's development from decades ago. For example, the period-profile in 1874 or in 1884 may reflect outcomes for people whose childhood years were in the 1840s and 1850s. Likewise, gains in children's situation in the 1870s and 1880s may not have swayed the rates of non-communicable disease until 1904 or 1914, when they first contributed at age 25–34, and then to subsequent period-profiles over next five to six decades. This type of intertemporal synergism could have been behind the three to four decade lag between the falling trend of infectious diseases and of the non-communicable ones in Chap. 2, Fig. 2.8.

Table 4.3 Summary of the relation between the simulated aggregate rates and the actual rates

	I	II	III	IV	All
Correlation	0.97	0.98	0.98	0.98	0.98
t-statistic of the correlation	14.4	18.4	17.8	18.4	38.2
Average percentage error	0.9%	1.2%	0.4%	1.0%	0.88%
Root mean squared percentage error	5.3%	5.2%	4.8%	5.2%	5.1%
Average error (rate per 100,000)	10.8	16.5	−1.7	−13.8	2.9
Root mean square error (rate per 100,000)	51.5	53.2	47.6	51.3	50.9
N	15	16	15	16	62

For each set, Table 4.3 summarizes how the simulated aggregate rates implied by the cohorts' α and β matched up to the actual aggregate rates. The first row shows that they correlate closely. In percentage terms, on average the simulated rates were about 1% off the actuals. The mean squared percentage error is less than 0.3%, and the root mean squared is about 4.8–5.2%. In terms of age-adjusted rates per 100,000, the simulated aggregate rate was off on average by 10.8 for set I, 16.5 for set II, −1.7 for III and −3.8 for IV. The root mean squared errors of the simulated rates were about 48–53 per 100,000, which is relatively small.

Further, all sets connected together yield a longer time series of the aggregate rates for which the results are shown in the last column. Figure 4.3 plots the simulated and the actual time series. By approximating the future period-profiles closely, the cohort parameters help track the direction and the level of the aggregate rate reasonably well. Figure 4.4 shows the percentage errors for the entire series. The mean square percentage error is 0.27%. It tended to be more during the two world wars, and in the 1970s than in other periods, though staying less than about 1% most of the time. The errors in the 1970s and 1980s could be reflecting the shortfall of the working assumption for the post-transition profiles, which was that they would fare no better than the 1920s profile at age 65–74 and no better than the 1930s profile at age 55–64.

The results indicate that the cohorts' α and β are a reasonably accurate guide to the aggregate rate's path. Chapter 8 explores the extent to which socio-economic influences from the childhood years had swayed the α and β across the cohorts. It first extracts the childhood-linked signals in these parameters, and then subjects the childhood-linked or the childhood-implied α and β to a similar test, which assesses how well they are able to simulate the actual period-profiles and the resulting time series. The departure from the actuals may then give an approximate size of the period-profiles' life-course component, which, by definition, is beyond the reach of childhood development. The results will be displayed in figures similar to Fig. 4.3.

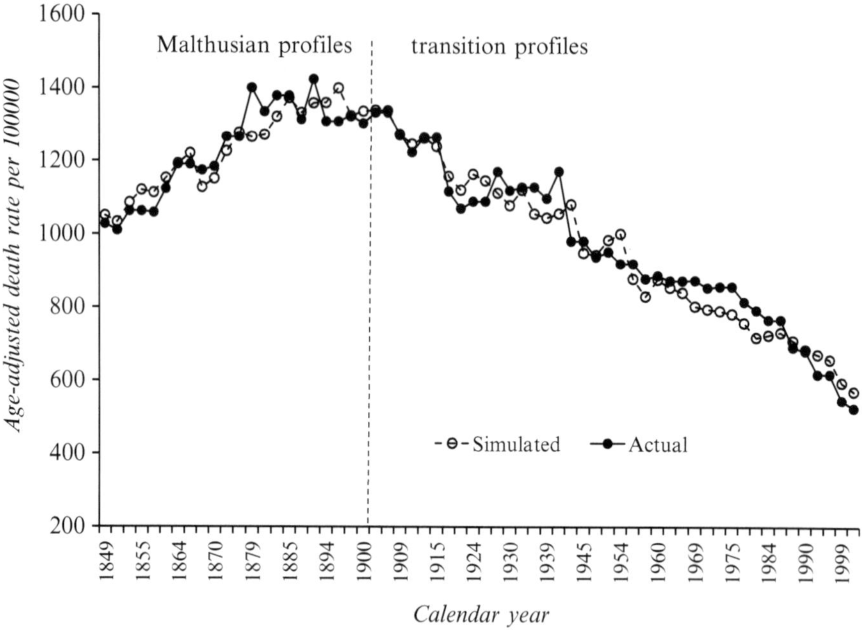

Fig. 4.3 The simulated rates versus actual aggregate rates of non-communicable diseases, various years, 1849–2000 (Sources: HMSO; author's calculations)

Fig. 4.4 The squared percentage errors between the simulated rates and the actual aggregate rates of non-communicable diseases, various years, 1849–2000 (Sources: HMSO; author's calculations)

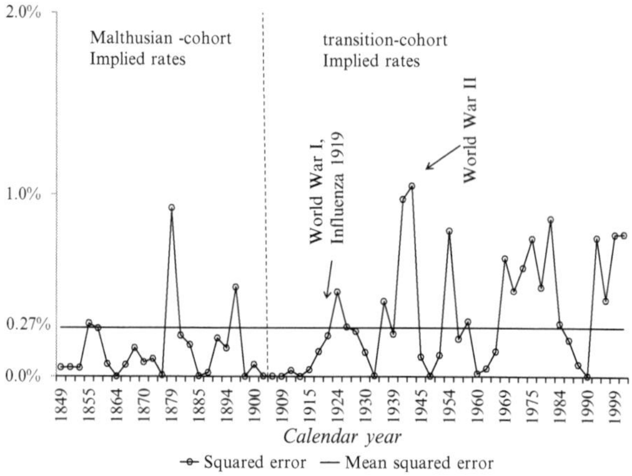

4.3 Which Exposes the Source of the Paradox

If non-communicable diseases climb with age (as all profiles affirm they do), then how is it that the aggregate rate has been falling over the twentieth century as average age was growing? The simulation indicates that the paradox can be resolved by recognizing the downward shifts in the cohort-profiles.

However, the complete set of options is that the aggregate can grow, shrink, or stagnate. The underpinnings for a growing or a shrinking aggregate rate are clear in Tables 4.1 and 4.2. Other than by chance during a period of inflection, under what type of situation could the aggregate stagnate over the long term? Such a situation may be only a theoretical one: there must be no shifts in the profiles at all; each generation would be identical and age along one universal upward-sloping stationary profile; and the death rates at each age of that profile would be unchanged no matter which generation is aging along it. The age-adjusted aggregate rate would then be the sum of the age-specific crude rates in the profile. The cohort- and the period-profile would coincide, and the aggregate rate would remain unchanged. In other words, α and β stay unchanged generation after another—the static profile.

4.4 And the Shortfalls of Period-Profiles

What the simulation shows instead is that the period-profiles had varied; were likely different from the cohort-profiles; and could have been swayed by cohort-level change. Subsumed in the cohort-level movements is the influence of medical technology at various times, and of various adverse or beneficial life-course events.

It should be clear from the simulation that any period-profile would have been an unreliable basis for framing an outlook of aging when the underlying conditions (epidemiologic, economic and political) were changing rapidly. This issue about the reliability of period-profiles is a critical practical matter. They are used for computing period-life expectancies, which, in turn, inform issues such as healthcare and other budgetary projections at various levels of government; for the pricing of annuities; for the evaluations of healthcare spending at the national level by multilateral agencies such as the ones conducted by the OECD, and are also included in measures of progress in human development such as the United Nations Human Development Index.

While period-life tables use deaths from all causes and in all age groups, the discussion below applies only to the extent that non-communicable diseases at ages 25–74 affect the calculation of life expectancy. Since the share of such diseases has risen from about 50% in the mid-nineteenth century to about 80–90% after World War II, the discussion below is likely to have become increasingly relevant.

One way to assess the reliability of period-profiles is to explore if they routinely overstate or understate aging potentials ahead, and then study the types of situations in which they err in either direction. That can be done in four steps: (1) estimate the parameters for the period-profiles; (2) calculate the implied period-aging potential (henceforth, period-potential) or the mortality rate doubling time in each period; (3) compare the period- and the cohort-potentials and calculate the gaps between them; (4) assess the underlying situations when the gaps appear.

However, one critical fact should be kept in mind: at any time, period-profiles are the most updated data available. The cohort-profiles, whereas, are unfinished until long afterward; and their data update only when the cohorts have aged enough. For that reason alone, period-profiles are of immediate practical value, and the information they provide should be taken very seriously. It is also correct to say that comparing them to cohort-profiles presumes that the data on cohorts would be available decades beforehand. In actuality that cannot happen. The two types of profiles can be compared only in retrospect, or with near-perfect hindsight. Nonetheless, there is some practical value in comparing them: it helps identify situations when the aging potentials they imply differ, and knowing the contexts could help frame likelier outlooks of aging.

Figure 4.5 shows the doubling times of cohort-profiles and of the period-profiles. Each cohort's doubling time is centered on the middle year of its birth decade. The doubling times of the period-profiles are centered on the decades when the deaths were reported; they are based on the average of age-specific death rates during the decade, reflecting the data an analyst would have by the end of the decade. For both types of profiles, a downward trend in the doubling times signals growing aging potentials, and an upward trend conveys shrinking potentials.

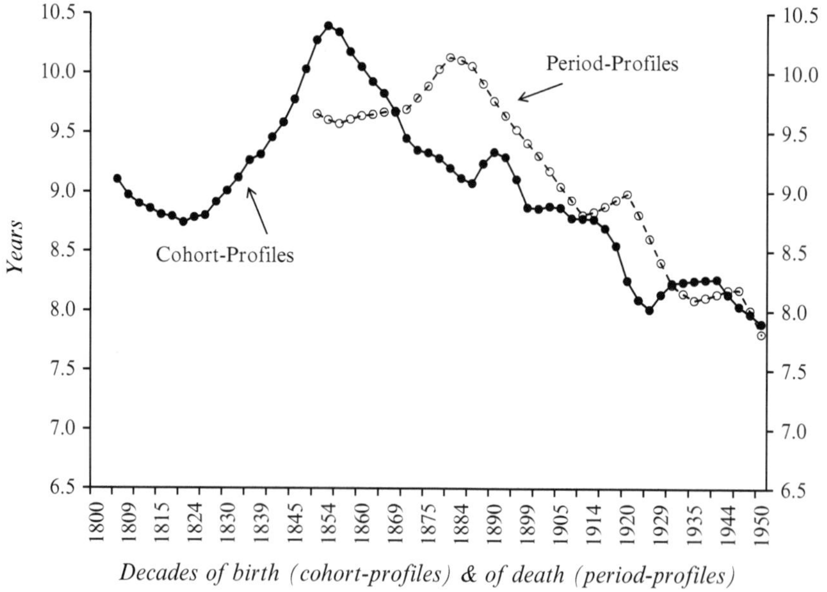

Fig. 4.5 Mortality rate doubling time of cohort-profiles versus period-profiles (4-year moving average), 1800–1950 (Sources: HMSO; author's calculations)

The cohort-potentials were first improving steadily in the early nineteenth century. They then reversed course for a series of cohorts raised through mid-century. After the reversal, successive cohorts have mostly fared better though there were some relatively mild reversals for the ones in the 1890s, and for the ones whose birth-years were in the inter-war period, especially during the Great Depression of the 1930s.

The period-potentials, meanwhile, seem to be echoing the trends in the cohort-potentials from nearly five decades earlier. The period-potentials in the 1840s and 1850s were echoing the cohort-potentials from the early nineteenth century as cohorts from then were the ones shaping the period-profiles in the 1840s and 1850s. The period-potentials reflected the turning for the worse as well, but did so since about mid-nineteenth through early twentieth century as the cohorts brought up since the 1820s began swaying the period-profiles. The period-potentials have since steadily improved except during the two world wars.

Moreover, in the mid-nineteenth century, period-potentials seemed too optimistic about the actual aging potentials of the cohorts who were children then. The reverse occurred over the last three decades of the century, when the period-potentials not only understated the cohort-potentials, but worsened just when children's stature was growing and infectious diseases were shrinking. The period-profiles from then neither had a way of knowing the intertemporal linkages that could show up later, nor the influence of any new technologies that may have become available in the decades ahead. Any outlook of aging based on those period-profiles is likely to have erred about the direction as well as the magnitude of the new cohorts' potentials.

Though the role of new medical know-how and other life-course changes is not directly visible, and should not be ruled out (Chaps. 8, 9, and 10 address the issue), a pattern that seems to show up consistently is that during times when children's status changes sizably, period-potentials tend to misgauge the eventual cohort-potentials. This seems to have been the case especially in the second half of the nineteenth century, where it is unlikely that the medical know-how was regressing and driving up the period-rates. Neither does it seem likely that people were becoming less educated than before. Implausible, too, is that the material living standard was falling over those 50 years of per capita real income growth. While none of them can be ruled out, it seems likelier that the period-profiles were degrading sequentially because successive cohorts contributing to the period-profiles were the ones that had endured setbacks in their childhoods decades earlier. The other possibility is that some other types of life-course events during that time were becoming consistently and systematically worse over 50 years.

Figures 4.6 and 4.7 clarify why there might be something to the conjecture about the connection with children's status. Figure 4.6 juxtaposes the initial states α_i of the cohort-profiles (shown as the line) and the gaps between cohort-potentials and period-potentials (depicted by the bars). The α_i express how well or poorly the cohorts began their adult life cycle: climbing α_i convey poorer beginnings, and most likely signal that the cohorts had endured more stresses as children. Meanwhile, the gap depicted by the bars is the doubling time of the cohort-profiles minus that of the

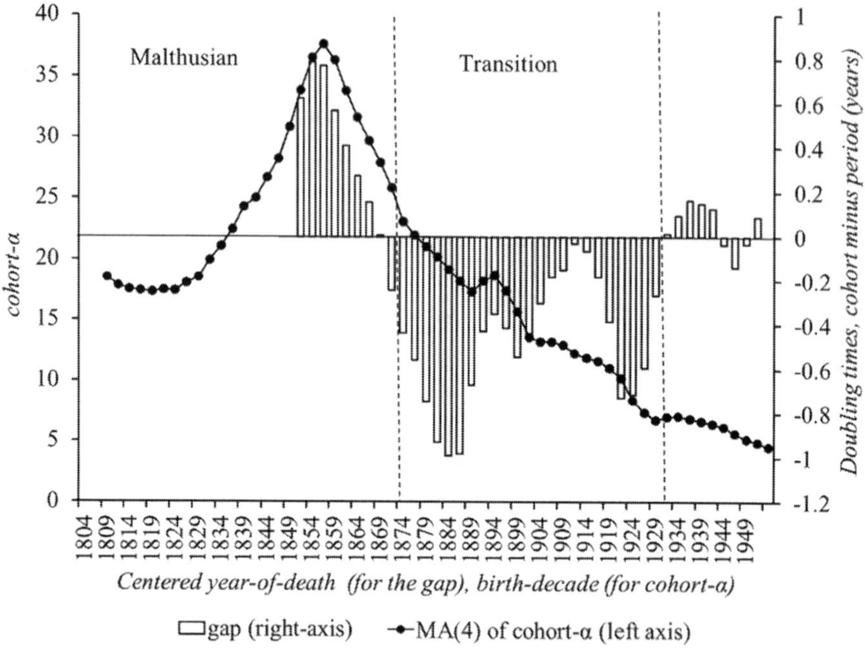

Fig. 4.6 The cohort-profile-α_i and the gaps between cohort- and period-potentials, 1800–1950 (Sources: HMSO; author's calculations)

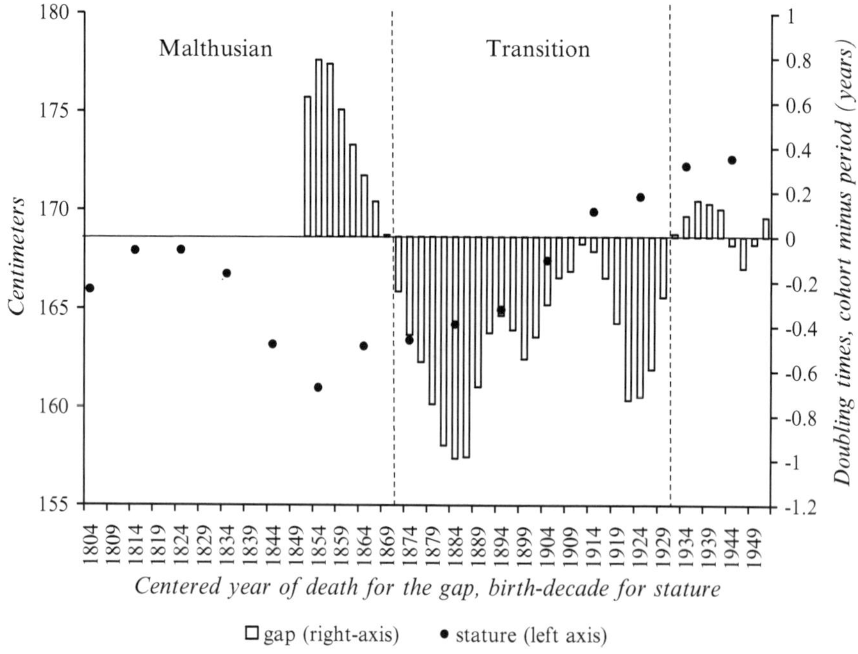

Fig. 4.7 Average adult stature of cohorts and the gaps between cohort- and period-doubling times, 1800–1950 (Sources: Floud et al. (1990), Rosenbaum (1988); HMSO; author's calculations)

period-profiles. A larger doubling time signifies poorer aging potential. So the cohort-doubling time minus the period-doubling time is positive when the cohort-potentials had turned out worse than period-potentials—or, when the period-potentials were too optimistic about the outlook of aging.

As successive cohorts of children had likely been faring worse during the early nineteenth century, the α_i of the cohort-profiles were climbing, and the gap between the cohort-doubling times and the period-doubling times was widening. The aging potentials implied by the period-profiles were too optimistic about the actual potentials for the cohorts who were children then. This same pattern shows up in Fig. 4.7: when children's physiological growth was suffering was also when the period-profiles were overstating the actual aging potentials. That pattern reversed in the transition stage: the period-profiles began understating the cohorts' potentials when progress in children's development was underway.

However, the gap became negligible during the 1910s. This happened as the doubling time of the period-profile for the decade grew because of the world war and the 1919 influenza pandemic (see Fig. 4.5). Meanwhile, the cohort-potentials were slightly larger than the period-potentials in the 1930s. In this instance, it was due to the cohorts brought up then; they fared worse than what the period-profiles of the time were indicating. The health-related ramifications of the Great Depression were felt less contemporaneously in the period-profiles, and more in the data for the cohorts who grew up then. This applies broadly to the inter-war cohorts.

The gaps between the doubling times raise the practical issue of how long may it take for the period-profiles to first reveal that something may have been amiss? This is critical for framing an outlook of aging because there is no way of completely knowing ahead of time how the cohorts may eventually fare. Life-course factors, by definition, are unforeseeable. At any point in time, the data actually available to an analyst is on partial cohort-profiles, and on the most recent period-profile. How soon could a period-profile signal that the burgeoning cohorts may be doing worse or better than the ones before?

The cohort-α_i and the period-α_i in Fig. 4.8 suggest that it may take at least three to five decades. The period-α_i in the nineteenth century did not turn decisively until about four decades after the cohort ones signaled a turning for the worse. Likewise, it may take period-profiles three to five decades to reveal with some confidence that any progress may have been underway.

It is important to be clear about the issue Fig. 4.8 illustrates. It is not that the cohort-α_i signal before the period-α_i. The cohort-α_i always will, because they are placed by the year of birth, and that too in retrospect, when the data on both types of profiles are available. The issue Fig. 4.8 clarifies instead is: if period-profiles were the only data available for finding out how the new cohorts might be faring, then how much time might it take to discover that a new trend had been underway (three to five decades), and which stage of the lifecycle may first reveal it (presumably the childhood years, as α_i refer to the initial states for the profile in adulthood). This issue will reappear in Chap. 10, which explores a list of things to inspect for analyzing burgeoning trends.

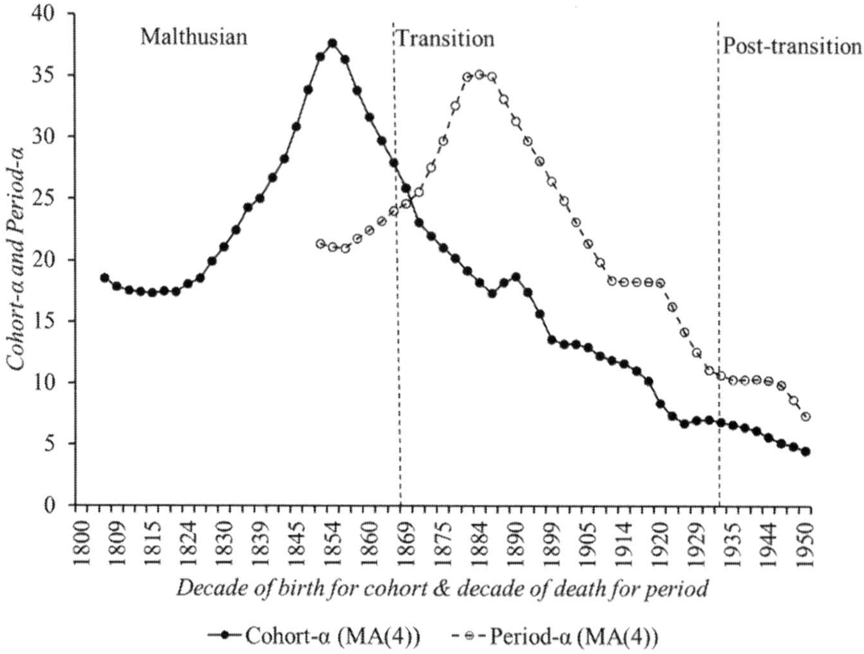

Fig. 4.8 Cohort-profile-α_i versus period-profile-α_i, 1800–1950 (Sources: HMSO; author's calculations)

4.5 Shedding More Light on the Outlooks of Aging

It helps at this point to reconnect with the debate on aging. As the received view's age of delayed deaths began in the 1960s and 1970s, the outlooks of aging became central to the debate on healthcare outlays in general, and to the outlays of old-age programs in particular. Under the rubric of 'delays,' the outlooks faced two very long-term issues: (1) period-life expectancies at older ages; (2) morbidity or illness rates at older ages. The delay in deaths meant that more people were qualifying for old-age programs and could remain in the programs longer than before. The *basis* on which period-life expectancies might keep growing became a critical matter.

The prospect of continuing delays, however, ran into the notion of compression. It implies that the delays at older ages can go on only so far because lives have to be finite. So period-life expectancies could grow, but the years gained, in absolute terms, would be much less at older ages, and would peter out at extreme old age or near some limiting age.

The upward slope of the period-profile of communicable diseases thus became central to the debate. Informing the debate was what is known as the Taeuber Paradox (Taeuber 1976; Keyfitz 1977). It developed out of the circumstances in the 1960s and the 1970s. Subduing, even eliminating, many infectious diseases by then had added years to period-life expectancies (which are not lifespan). The question was: what might happen to the expectancies if similar success were achieved against many non-communicable diseases, especially ones whose rates were still elevated then? Subduing infectious diseases had let people live longer; subduing non-communicable ones would add even more years.

While it was obvious that the outcome would be more delays, but that would also make people more prone to degenerative conditions, which tend to intensify with age. Seen from the vantage point of *within* any profile, or a static profile, the delays should then prove transient because all profiles slope upward. As a result, period-life expectancies, especially at older ages, might not have too much room to grow.

However, in light of the evidence on the cohort-profiles, one could state the outcome somewhat differently. Cohort-profiles, too, slope upward; newer profiles are even steeper than older ones. All of them are subject to Taeuber's paradox. However, as was clear in the simulation, several unique cohort-profiles shape each period-profile. If one presumes that non-communicable disease in ages 25–74 were the only input into period-life tables, then just like the period-profiles had improved in response to an improving cohort mix, period-life expectancies may also grow as the cohort mix tips in a favorable direction. However, since cohort-profiles tend to converge at advanced ages, all else equal, the absolute gains—not percentage gains—in older age period-life expectancies may not turn out to be very big, or at least not as big as the ones at younger ages.

Further, just as the period-potentials had misgauged cohort-potentials, period-life expectancies could misgauge cohort-life expectancies. That may happen for at least two reasons: (1) a new medical technology may appear in the future (Goldman et al. 2005); (2) children's development, and the socio-economic factors that affect it, may change. All else equal, period-life expectancies could grow if childhood development of newer cohorts is better, and they start out on unique profiles that converge from below toward the profiles of preceding cohorts. The reverse could occur if the new cohorts of children fare worse than the ones before (Olshansky et al. 2005).

Moreover, besides misgauging prospects, period-life expectancies could mislead inference about the *current* period as well. For example, life expectancy at age 20 grew decisively from about 40 years in 1850/1870 to about 48 years in 1920, likely because infectious diseases were falling then. However, one would not know from the gain alone that rates of non-communicable diseases, in ages as early as 25–34, had been growing for about 50 years. The issue here is that the mix of causes is irrelevant to the calculation of period-life expectancy; it just sums up the outcome of all deaths. Its growth is usually inferred as progress against deaths, if not as advances in health. However, when trends in diseases that make up at least 50% of deaths differ from the trend of period-life expectancy, as was the case in the second half of the nineteenth century, the usual inferences about health or wellbeing based on life expectancy alone are unsafe.

Aside from relating to period-life expectancies, the period- and a cohort dimensions extend to the notion of the compression of deaths. Over the years (periods), deaths at younger ages have fallen; successive period-profiles have shifted down, become steeper, and indicated better period-potentials. For period- as well as cohort-profiles, all else constant, steeper slopes and shrinking doubling times signify compression; flatter slopes and growing doubling times convey expansion.

However, they differ in one important way. As Fig. 4.5 shows, trends in the period-profiles are shaped by several contributing cohorts that may tug in opposing directions, especially at turning points. The aging baby-boom cohort, for instance, will not be the only one shaping the period-profiles in the future.

The second main element in the debate about the prospects of aging is the compression of morbidity or illnesses. The delays in deaths clearly mean longer lives, but they leave open the issue of whether the lives are healthier than before. Scholars have framed this issue in terms of three possibilities for the 'healthy years' as a *share* of the lifespan: the share of healthy years (1) stays unchanged if the prevalence of non-communicable diseases and the gains in person-years advance proportionally; (2) falls if prevalence grows more than the person-years gained; (3) and rises if prevalence either shrinks or grows less than the gain in person-years. The compression of morbidity refers to the third scenario. All else constant, healthcare outlays grow with age under all of them, but they grow the slowest in the compression scenario as the demand for healthcare from the oldest of the old grows less than in the other two. This is the basis for the 'healthy aging' scenario and the bend in the spending profile in Chap. 1.

While there is clear evidence on the trends in compression of *deaths* over the past two centuries, long-term evidence on the trends in morbidity, let alone its compression or expansion, is fragmented and sparse. Chapter 10 will return to this issue while modeling illness and aging profiles in relation to healthcare spending.

Meanwhile, the next chapter studies if the cohort-level trends in infectious diseases were similar to the ones in non-communicable diseases. Excepting for a mild case for respiratory infections, they were not. The main takeaway is that subduing infectious diseases had an immediate effect on aging: people gained more years right away. However, the ramifications of subduing infections may have extended inter-temporally as well: cohorts of children that grew up in conditions increasingly safer from infections may have aged more than the ones that did not.

To go directly to the exploration of the intertemporal channel, the reader may want to skip the next chapter and start with Chap. 6. It explores the economic and political milieus the cohorts likely received in their childhood years.

References

Floud, R., Wachter, K., & Gregory, A. (1990). *Height, health, and history: Nutritional status in the United Kingdom, 1750–1980*. Cambridge: Cambridge University Press.

Goldman, D., et al. (2005, September). Consequences of health trends and medical innovation for the future elderly. *Health Affairs, 24*(Suppl 2), W5R5–W5R17.

Keyfitz, N. (1977). What would it make if cancer were eradicated. An examination of the Taeuber paradox. *Demography, 14*(4), 411–418.

Olshansky, J. B., et al. (2005). A potential decline in life expectancy in the United States in the 21st century. *New England Journal of Medicine, 352*, 1138–1145.

Rosenbaum, S. (1988). 100 years of heights and weights. *Journal of the Royal Statistical Society, 151*, 276–309.

Taeuber, C. (1976). If nobody dies of cancer. *Kennedy Institute Quarterly Report, 2*(2), 6–9.

Chapter 5
Most Infectious Diseases Diminished with Age

The 'representative' profiles of non-communicable diseases in Chap. 3 displayed three main features of aging. They slope upward with age, signifying that the rates climb as people get older. They had shifted downward as early as age 25–34 and had become steeper, suggesting that aging potentials had grown. They had shifted in a non-parallel way, and their tendency to converge at advanced ages suggested that cross-cohort gaps narrow with age. This chapter explores if the profiles of infectious diseases display those features.

The main difficulty in studying the infectious group at the cohort level is that the diseases included in it are diverse. Unlike most non-communicable categories that are grouped by specific anatomical site (the sub-categories), the infectious category has diseases that can pertain to many different anatomical sites. For instance, diarrhea and cholera affect the digestive system; encephalitis, meningitis and polio strike the nervous system; influenza and pneumonia, the respiratory system; rheumatic fever and aortic syphilis, the circulatory system.

Moreover, infectious diseases can be acute (e.g. influenza), or can last months or years (e.g. tuberculosis). Their diverse etiologies can also overlap. For example, respiratory tuberculosis, influenza, and pneumonia, among others, are air-borne; some types of tuberculosis and food poisoning are food-borne; diarrheal diseases, gastrointestinal infections, and cholera are water-borne. Consequently, changes to, say, food or water supply, can influence the trends in many such diseases together. Furthermore, their trends can be gradual, relating to the nutritional status of the population, or to the dissemination of preventive practices; and yet, sometimes, the trends can be episodic, like the various outbreaks, often triggered by mutated microbes (e.g. flu viruses).

Nonetheless, they all share the common trait of rapidly affecting a large number of people, and for that reason they have been subject to epidemiologic monitoring, even before it was legislated in the Infectious Disease Notification Act 1889. This rationale for monitoring them was clear to the designers of early ICDs that had grouped many such diseases under the miasmatic category (see Appendix). To simplify matters, the chapter examines cohort-level trends in three main sub-groups

S. Arora, *The Transitions of Aging*, International Perspectives on Aging 12,
DOI 10.1007/978-3-319-14403-0_5

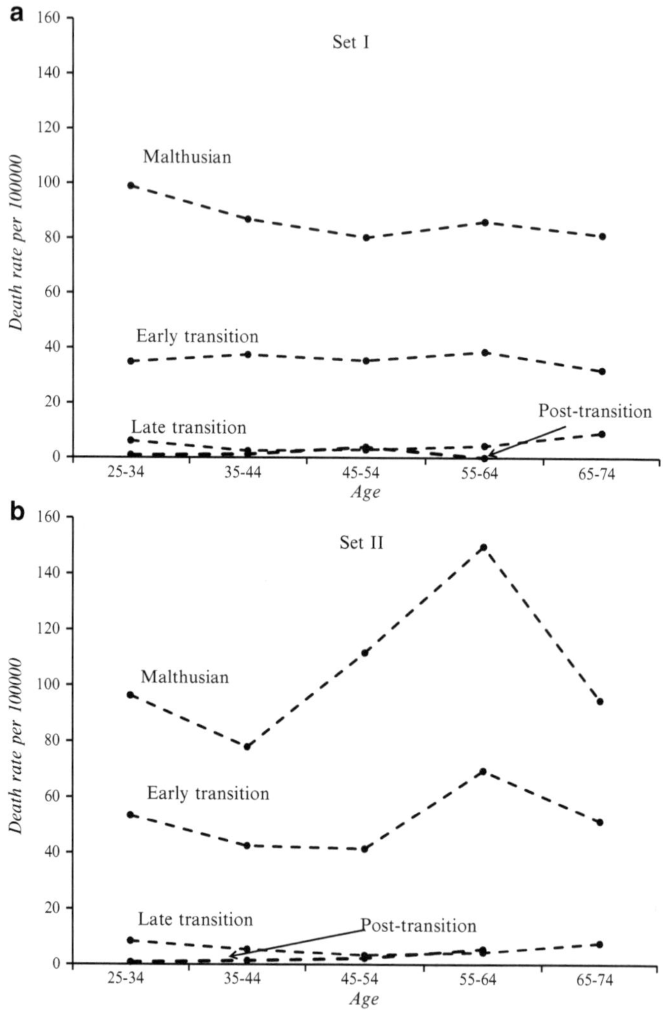

Fig. 5.1 Age-profiles of infectious and parasitic diseases excluding tuberculosis (**a**) Set I (**b**) Set II (Sources: HMSO; author's calculations)

of infectious diseases: (1) All infectious diseases excluding tuberculosis; (2) tuberculosis; (3) respiratory diseases, which are dominated by influenza, pneumonia and bronchitis.

Figure 5.1a, b show for sets I and II the 'representative' age-profiles of all infectious diseases excluding tuberculosis. The profiles do not slope upward with age. Unlike the profiles of non-communicable diseases, they differ markedly across the sets. For instance the Malthusian profiles and the early-transition profiles between set I and set II suggest that period-level developments instead of an underlying age-specific intrinsic force, such as senescence, may have shaped each profile.

The profiles also jumped sizably from one regime to the next. The gaps between the Malthusian, the early-transition and the late-transition profiles at age 25–34 reflect the large advances made against infectious diseases between the 1870s/1880s and the 1910s/1920s, and between the 1910s/1920s and the 1940s/1950s. Further, the gaps between the profiles have narrowed as the transition proceeded. That too is resonant of the period-rates in the transition stage. In the initial stages, the incremental progress was large, and the gaps are wider. As increasingly more diseases were subdued, the size of the marginal progress diminished. By the time the post-transition profiles got started, they display only minimal progress relative to the late-transition profile's start, as well as to the rest of its profile. Further, the nearly negligible rates in the post-transition profiles suggest that the progress achieved by the 1930s and 1940s was sustained through the World War II era. All of these features seem to echo period-level developments, and may have been driven by common factors such as better public health infrastructure; adoption of new knowledge by medical practitioners and households; and the advent of antibiotics; among others.

The profiles of tuberculosis in Fig. 5.2a, b are different from the ones of other infectious diseases. Though they too show large decreases in the starting points, likely echoing period-level developments, the Malthusian and the early-transition profiles clearly display that rates of tuberculosis had decreased with age. One reason could be better resistance among those who had aged more than others. Tuberculosis, especially respiratory tuberculosis, later in life has been known to be usually caused by an infection acquired well in the past, and its incidence at older ages was much less than at younger ones (Brownlee 1916; Frost 1939; Springett 1951, 1952; Lowe 1954; Spicer 1954). Further, as each cohort had aged, it had faced increasingly better overall epidemiologic environment, especially as the spread of infectious diseases was reduced after better surveillance since the Notification Act 1889.

The likely role of another such period-effect on the cohort-profiles shows up in the comparison of the late-transition and the post-transition profiles in both sets. At age 35 and 45, the late-transition profile collapsed toward the horizontal axis. That occurred when chemotherapy and the BCG vaccine, along with subsequent advances in the treatment of tuberculosis—all period or life-course developments—became available to the post-transition cohort throughout its life cycle. Its age-profile became nearly indistinguishable from the horizontal axis. Some of the sequential change in slopes of the other profiles is related to the advent of the new treatments as well. For example, the early-transition profiles (born roughly 1865–1895) were about 50–60-years old when the new treatments arrived, and that is also when their profiles descended faster than before.

Moreover, tuberculosis is widely seen as a nutritionally-related disease. The late- and the early-transition profiles pertain to people who were brought up when per capita income was nearly twice the level when people in the Malthusian profile were raised. That change may have influenced some of the gaps between their profiles, at least until when the new treatments arrived. Further, it should be kept in mind that the relationship between infectious diseases and nutritional status is

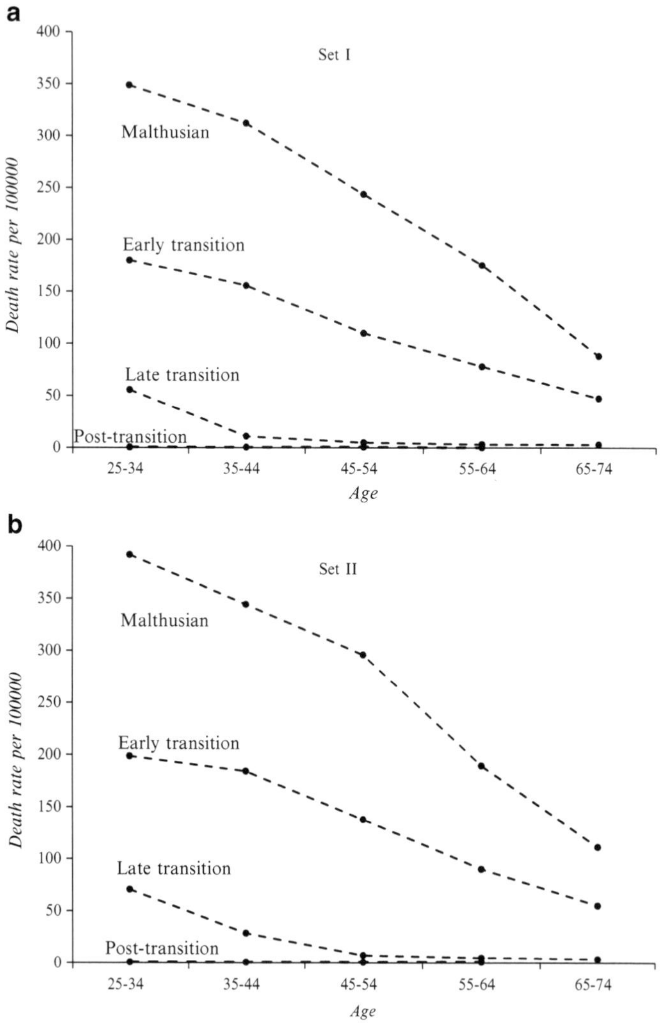

Fig. 5.2 Age-profiles of tuberculosis (**a**) Set I (**b**) Set II (Sources: HMSO; author's calculations)

synergistic; and the incidence of infections itself can undermine the nutritional status (Martorell et al. 1975; Mata 1978; Martorall 1980; Martorell and Habicht 1986; Dasgupta 1993; Voth and Leunig 1996). In that vein, containing other infectious diseases could have boosted nutritional status, and may have indirectly helped widen the gaps between the profiles.

Above all, the profiles provide a unique view of how new technology, a period-event, may affect cohort-level patterns of diseases that may not be linked to aging in the same way as various non-communicable diseases are linked. The behavior of the late- and the post-transition profiles indicates that a new effective technology

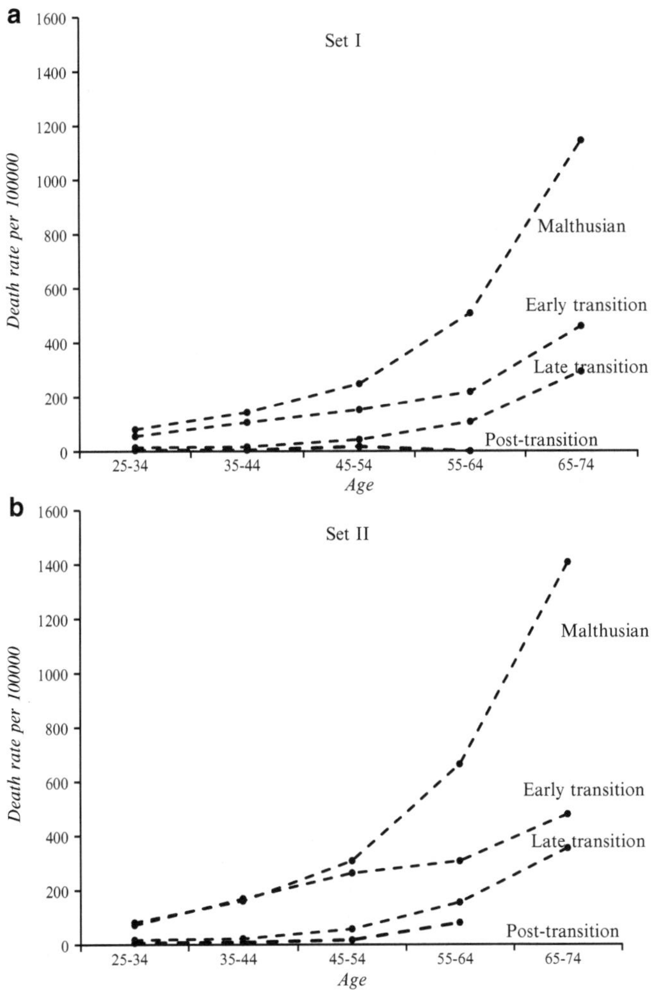

Fig. 5.3 Age-profiles of respiratory diseases (**a**) Set I (**b**) Set II (Sources: HMSO; author's calculations)

can collapse the profiles rapidly, resulting in lasting benefits for existing as well as future generations. While the time series view of crude death rates studied in Mckeown (1976) had suggested a secondary role for new technology or for public health infrastructure, the generational view emphasizes that their roles may be better assessed by what has occurred to the generational profiles, which show rapid and lasting benefit.

The profiles of respiratory diseases in Fig. 5.3a, b seem to display a somewhat different, but even more mixed picture than the other two groups. The upward age-coincidence is clear in the Malthusian profiles and their respiratory infections had increased exponentially with age. However, the other profiles show either very subdued age-coincidence or none at all. For example the post-transition profile for set I is nearly indistinguishable from the horizontal axis. Further, in set II, the early-transition profile overlapped with the Malthusian one until about age 50 before tracing a path of its own. Antimicrobial therapies had become available to the early- and late-transition cohorts by the time they were age 65–74, but were unavailable at the time the Malthusian cohort was 65–74 years old. While time-specific events may have played a role, several micro-level studies of respiratory diseases in adults have implicated the role of childhood infections as well (Wesley 1991; Barker and Osmond 1986a, b; Barker et al. 1991; Shaheen et al. 1994; Shaheen 1997; Mazumdar et al. 2010).

All told, except for some respiratory infections, the profiles of the infectious group do not display the three main features seen in the profiles of non-communicable diseases. However, the fall in infections was of immediate consequence to aging: more people were able to become older aged. Further, the time-trends of infectious diseases outlined in Chap. 2 indicated that successive cohorts of children were being raised under increasingly better epidemiologic conditions. That could have influenced the initial states and the slopes of their profiles of non-communicable diseases. If true, then restraining infectious diseases may have boosted aging indirectly as well. Chapter 8 will explore this indirect channel.

References

Barker, D. J. P., & Osmond, C. (1986a). Infant mortality, childhood nutrition and ischemic heart disease in England and Wales. *Lancet, 1*(8489), 1077–1081.

Barker, D. J. P., & Osmond, C. (1986b). Childhood respiratory infections and adult chronic bronchitis in England and Wales. *British Medical Journal, 293*(6557), 1271–1275.

Barker, D. J. P., Godfrey, K. M., Fall, C., Osmond, C., Winter, F. D., & Shaheen, S. O. (1991). Relation of birth weight and childhood respiratory infection to adult lung function and death from chronic obstructive airways disease. *British Medical Journal, 303*(6804), 671–675.

Brownlee, J. (1916). Certain conditions regarding the epidemiology of phthisis pulmonalis. *Public Health, 29*, 130–145.

Dasgupta, P. (1993). *An inquiry into well-being and destitution*. Oxford: Clarendon Press.

Frost, W. H. (1939). The age selection of mortality from tuberculosis in successive decades. *American Journal of Epidemiology, 30*(section A (3)), 91–96.

Lowe, C. R. (1954). Recent trends in survival of patients with respiratory tuberculosis. *British Journal of Preventive and Social Medicine, 8*(3), 91–98.

Martorell, R. (1980). Inter-relationships between diet, infectious disease and nutritional status. In L. S. Greene & F. E. Johnsston (Eds.), *Social and biological predictors of nutritional status, physical growth and neurological development*. New York: New York Academic Press.

Martorell, R., & Habicht, J. P. (1986). Growth in early-childhood in developing countries. In F. Falkner & J. M. Tanner (Eds.), *Human growth: A comprehensive treatise* (Vol. 3). New York: Plenum.

Martorell, R., Habicht, J. P., Yarborough, C., Lechtig, A., Klein, R. E., & Western, D. (1975). Acute morbidity and physical growth in rural Guatemalan children. *American Journal of Diseases of Children, 129*(11), 1296–1301.

Mata, L. J. (1978). *The children of Santa Maria Cauque: A prospective field study of health and growth*. Cambridge, MA: MIT Press.

Mazumdar, B., Almond, D., Parka, K., Crimmins, E. M., & Finch, C. E. (2010). Lingering prenatal effects of the 1918 influenza pandemic on cardiovascular diseases. *Journal of Developmental Origins of Health and Disease, 1*(1), 26–34.

Mckeown, T. (1976). *The modern rise of population*. London: Edward Arnold.

Shaheen, S. O. (1997). The beginnings of chronic airflow obstruction. *British Medical Bulletin, 53*(1), 58–70.

Shaheen, S. O., et al. (1994). The relationship between pneumonia in early childhood and impaired lung function in late adult life. *American Journal of Respiratory Critical Care Medicine, 149*(3), 616–619.

Spicer, C. C. (1954). The generation method of analysis applied to mortality from respiratory tuberculosis. *Journal of Hygiene, 52*(3), 361–368.

Springett, V. H. (1951). Results of reexamination of mass radiography. *British Medical Journal, 2*(4724), 144–148.

Springett, V. H. (1952). An interpretation of the statistical trends in tuberculosis. *Lancet, 259*(6708), 521–525.

Voth, H., & Leunig, T. (1996). Did small-pox reduce adult height? Stature and the standard of living in London, 1770–1873. *Economic History Review, XLIX*(3), 541–560.

Wesley, A. G. (1991). Prolonged after-effects of pneumonia in children. *South African Medical Journal, 79*(2), 73–76.

Chapter 6
The Macro Backdrop for Children

6.1 Was of Economic Growth

As a pioneer of modern economic growth, defined as a sustained rise in real per capita income over the long term, the U.K. continues to be in the club of rich countries today. However, as Table 6.1 shows, economic growth was uneven over the decades since the early nineteenth century. Real per capita incomes grew relatively slowly during the classic years Industrial Revolution (Crafts 1985a, b; 1994), picked up pace mid-century, but gradually slowed again during the second half, which has been the source of a longstanding puzzle (McCloskey 1970 "Did Victorian Britain Fail?"). Economic growth was quite volatile in the inter-war years that included the Great Depression, though in Table 6.1 the average rates in those decades hide the large intra-decade variation. After world war II, per capita incomes grew robustly, surpassing 2% average annual rate in most decades.

Recent research on economic growth has begun exploring intertemporal effects of childhood health. It examines non-communicable diseases that appear years later, and the consequences for productivity and economic growth (e.g. Aksan and Chakraborty 2012; Howitt 2005). This line of work may relate closely to the literature on slower productivity outlined in McCloskey (1970), as the evidence so far in this book appears to suggest that setbacks to childhood development in the mid-nineteenth century may have reached long into the second half of the nineteenth century when non-communicable diseases among adults were climbing.

Despite unevenness in the pace of economic growth over the decades, successive generations of children were brought up under increasing prosperity over the long term. Figure 6.1 shows for each decade the average *level* of real per capita incomes on the left-hand side scale. Compared to children who grew up in the early nineteenth century, real per capita income was 10% higher for the early-transition children; had doubled for the mid-transition children; and had more than tripled for the children in the late-transition and post-transition stages.

S. Arora, *The Transitions of Aging*, International Perspectives on Aging 12,
DOI 10.1007/978-3-319-14403-0_6

Table 6.1 The macroeconomic backdrop to the transition, 1800–1950

	Real per capita GDP growth (%)	Fiscal debt (% GDP)	Inflation (%)	Fiscal balance (% GDP)	Consol yield (%)	Return on Pvt. capital (%)
1801–1810	0.5		1.2		4.8	5.5
1811–1820	−0.4		−1.7		4.6	5.8
1821–1830	0.8	152	−1.2		3.6	6.0
1831–1840	1.3	156	0.1	−0.4	3.4	6.1
1841–1850	−0.6	145	−2.5	0.1	3.2	6.2
1851–1860	2.0	111	1.3	−0.4	3.2	6.3
1861–1870	1.2	78	0.2	−0.1	3.3	6.1
1871–1880	0.9	62	−0.1	0.0	3.2	5.5
1881–1890	1.4	60	−0.6	0.1	2.8	4.9
1891–1900	1.2	48	0.5	−0.3	2.4	5.2
1901–1910	0.3	39	0.4	−0.2	2.9	5.3
1911–1920	−2.2	84	10.6	−18.3	4.0	7.4
1921–1930	1.9	170	−3.6	1.3	4.5	6.9
1931–1940	1.8	185	1.7	−3.6	3.4	10.6
1941–1950	−1.5	220	5.0	−12.1	3.1	11.1

Sources: Fiscal balance, Inflation and consol yields, Hills and Thomas www.bankofengland.co.uk/publications; returns on capital (Piketty 2014); Real per capita income growth Maddison (2001)
Note: Rounded off averages of the annual rates in each decade. Inflation is for consumer prices; yields are for long-term Consols; private capital returns are average at the end of each decade. The fiscal balances for 1931–1940 and 1941–1950 excluding the War years are 0.4% and 3.5%

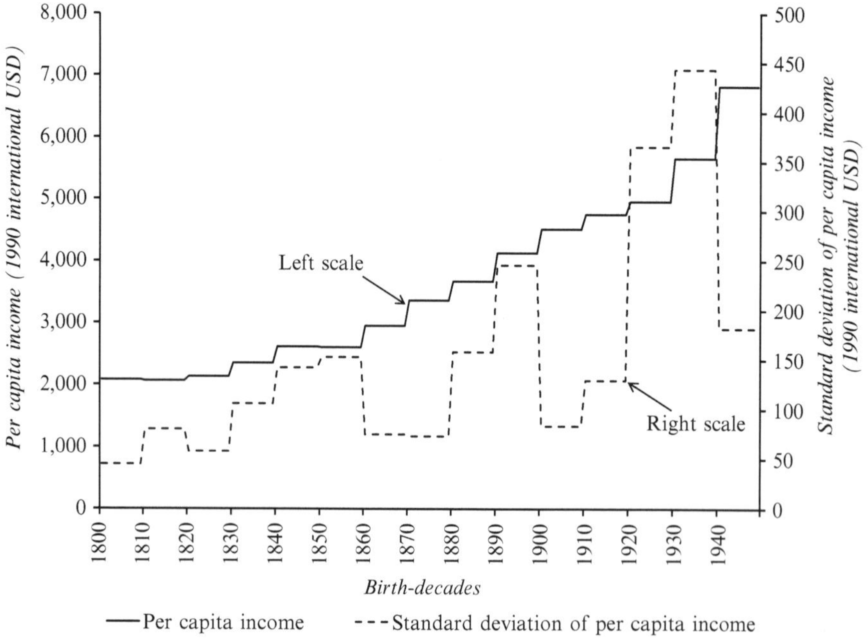

Fig. 6.1 Level of per capita income and its volatility, various decades, 1800–1950 (Sources: Maddison (2001); author's calculations)

While incomes had grown across the decades, the volatility of incomes had varied sizably. The right-hand scale of Fig. 6.1 shows the standard deviation of real per capita income in each decade. Volatility grew toward the mid-nineteenth century compared to the early nineteenth century; picked up again for the early-transition children; and was the highest for children who grew up in the inter-war years. This unsteadiness of incomes may have been critical. Though the volatility itself may have been short-lived, the dislocation accompanying it could have had lasting effects on the age-profiles. This issue will become clearer below in the discussion on stature, where there is a strong possibility of hysteresis-like consequences to aging. Chapter 8 will use the volatility-adjusted real per capita income in the childhood years as one of the macro variables alongside other childhood variables.

6.2 Together with Severe Fiscal Constraints

Economic growth in the nineteenth century was accompanied by a very high level of sovereign debt (Fig. 6.2). Financed primarily by private savings, the debt had built up to fight the American Revolution and the Napoleonic Wars. Debt service as a percentage of government spending grew from 39.1% in 1790

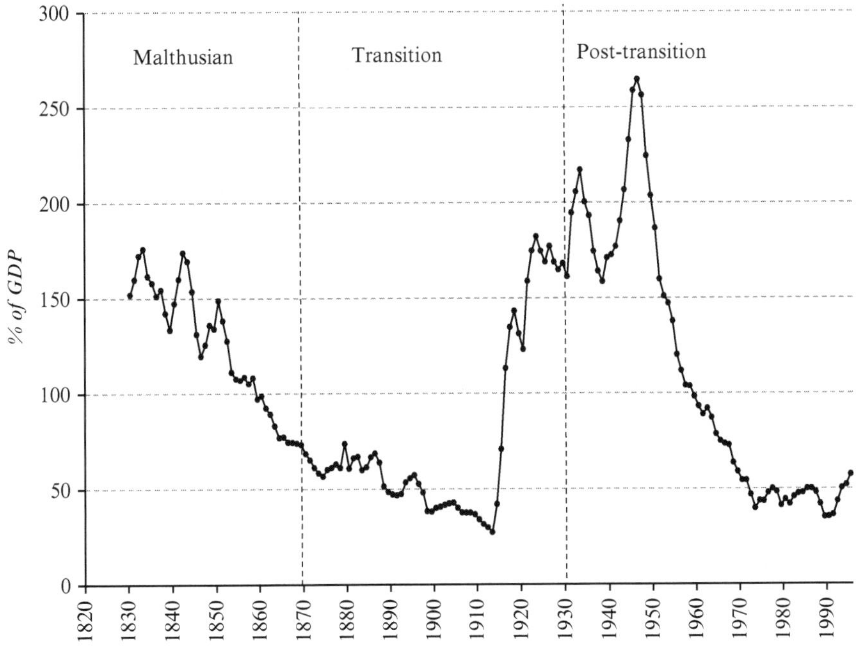

Fig. 6.2 Sovereign debt since the early-nineteenth century (Source: Bank of England (www.bankofengland.com.uk/publications))

to 42.2% in 1840, and shrank very slowly to 7% through 1900 (Harris 2004a, Table 1.2, p. 12). It took much of the century to draw debt down to less than 50% of GDP (Table 6.1).

Sovereigns usually draw down debt in three ways: (1) default or repudiation, (2) inflation, and (3) repayment. At no point there seems to have been an outright default. When debt levels are high, the temptation to inflate it away is usually strong as inflation can cut the real debt burden. However, the monetary regime of the gold standard meant that the inflation option was not used over the long term. Table 6.1 shows that most decades either experienced deflation or stable prices, and the average inflation rate was just above 1% in only two decades (the first one at least partially because convertibility to gold was temporarily suspended during the Napoleonic Wars).

The fourth column indicates that debt was likely repaid by adopting an austere fiscal policy. The fiscal deficit in each decade was very small. There were budget surpluses in about 50 out of the 70 years 1830–1910. This mode of debt repayment may have acted as a constraint to early industrialization, which is usually the time when the masses migrate to industrial towns, and the need for public goods and infrastructure is urgent (Williamson 1990). The severity of the epidemiologic picture discussed in Chap. 2 suggests that authorities may not have invested enough in the infrastructure to alleviate the situation, especially in the rapidly-urbanizing areas. In contrast to the decades-long austere fiscal policy followed to pay down the debt in the nineteenth century, the debt accumulated to finance the two world wars was drawn down more rapidly by a combination of inflation and robust economic growth after World War II.

Besides making it challenging to address the epidemiologic situation, fiscal austerity in the nineteenth century may have directly worsened the conditions for working-class children as the Poor Laws were diluted in 1834. The first half of the nineteenth century had coincided with an increase in child labor, which exposed children to an increasingly hazardous environment. Jane Humphries (2013) points out that "child labor was no anachronism inherited from a more brutal past. Instead, it looks to have been reinvented and propagated in the crucible of industrialization" (p. 177). In her study of childhood and child labor, average age when children started work fell from about 10.3 years in 1791–1820 to 9.98 years in 1821–1850, and subsequently rose to 11.4 years in 1851–1878 (Humphries 2013, Table 7.1, p. 177). The participation rates of children followed a similar pattern, being the highest in the period when the average age of the child laborers had been the lowest (Horrell and Humphries 1995).

Working-class children thus may have faced the consequences of a severe epidemiologic situation as well as an austere fiscal policy. "Neither the Old nor the New Poor Law was soft on the children of the working-class. The Old Poor Law was probably more generous in the provision of occasional doles, and through pauper apprenticeships may have provided lifelines for impoverished children. The New Poor Law was probably harsher in insisting on institutionalization as a precondition for relief but it did not provide workhouse schooling... The destiny of working children was to work and the offices of poor relief were to promote industrious independence" (pp. 45–46); and there is "ample evidence of pressure on poor families to employ their children at young ages, including the withholding of relief until children were so employed" (p. 194), with adverse consequences for human capital formation and inter-generational propagation of poverty (Horrell et al. 2001).

6.3 And of Vast Economic Inequality

The austere fiscal policies in the nineteenth century, meanwhile, coincided with vast inequality of income and wealth. Piketty (2014) has argued that in times when the rate of return on capital exceeds the rate of economic growth, a condition true until the first half of the twentieth century, accumulated private wealth intensifies economic disparity as wealth recapitalizes faster than the growth rate of the economy. The gap between the rate of return and the economic growth rate acts as a strong multiplier for capital and wealth, creating an incentive to live on capital incomes. These include rents, and fixed income returns such as the ones on government bonds, whose market values are protected by austere fiscal policy that reins in fiscal deficits, and by tight monetary policy that checks inflation. For instance, he points out that if the economy were growing at 1% annual rate and if the return on capital were 5%—roughly the situation in the early stages of the Industrial Revolution—savings of merely 20% of capital income helps capital grow faster than the rate of economic growth, boosting accumulated wealth further. Wider the gap between the two rates, more favorable the arithmetic is for an inheritance-based social set up, like the one of the propertied elite so vividly portrayed in Jane Austen novels.

Boosted by the multiplier, private capital had reached 750–800% of GDP in the first half of the nineteenth century, and had stayed near 700% of GDP toward the end of the century (Piketty 2014). In the 1850s, for example, rent alone was about 14–15% of national income, a share not seen since (Mitchell 2011). In contrast, by the 1950s and 1960s, private capital's share of GDP had shrunk to 300–350% of GDP after economic growth had outpaced the rate of return on capital for several decades, and mainly because the destruction over the inter-war years had cut the rate of return on capital sizably. By this time, rent, for instance, had shrunk to about 5% of national income.

The gap between the rate of return and the rate of economic growth tends to lift the share of capital incomes in the functional distribution of national income. Figure 6.3 shows that capital's share trended up, peaking at about 43% of GDP circa 1860. Its mirror image is the slide in labor's share, which gives an aggregate-level view of the long-running debate on the real wages of workers in various sectors of the economy (Lindert and Williamson 1983, 1985a, b; Crafts 1985a, b, 1989; von Tunzelman 1979; see Allen 2009 for the most recent statement).

The long-term trends in the functional distribution coincided with extreme disparity of wealth across the social classes, especially in the nineteenth century.

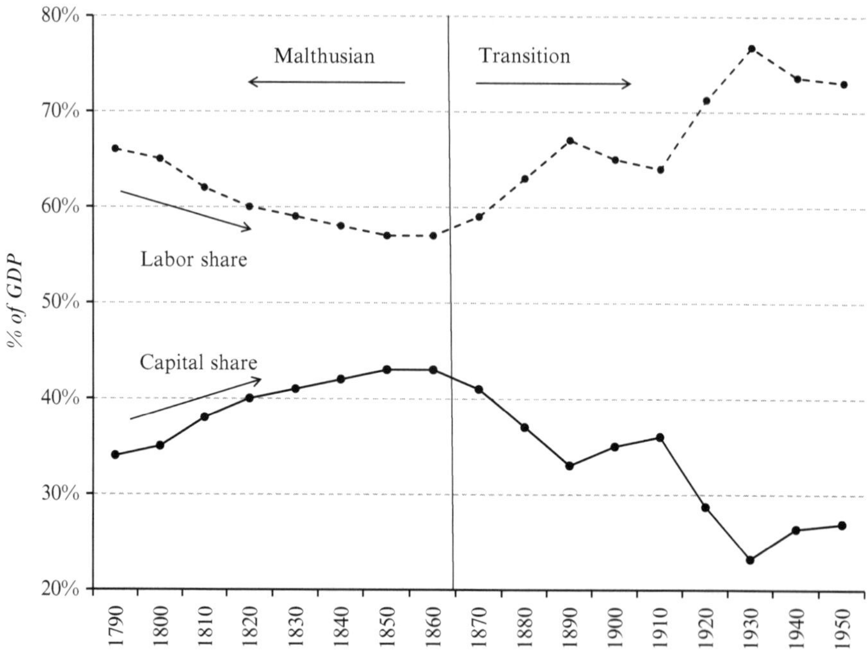

Fig. 6.3 The functional distribution of national income, 1790–1950 (Sources: Allen (2009) through 1910; Bank of England, thereafter (www.bankofengland.com.uk/publications))

Lindert (2000) estimates that 85% of the personal net worth in England Wales, as late as circa 1875, was held by the top 5% of wealthy adults. Their share of national wealth stayed around 85% through 1900, fell to about 80% in the first two decades of the twentieth century, dropping further to roughly 60% circa 1950. Based on probate data, scholars have found that the gaps in wealth narrowed from the 1910s through the 1950s and 1960s, and that coincided with falling shares in national income of the top 1%, 5% and 10% of households (Atkinson and Piketty 2007, 2010; Atkinson et al. 2011).

Like the inequality of wealth, income inequality widened over 1740–1810, fell modestly but stayed very wide over 1860–1910s, and then narrowed over the next several decades. In England and Wales, the top 20% of households received 55–60% of pre-tax income circa 1750, and their share rose to about 60–65% circa 1800, falling back to a still-sizable 55–60% in the 1860s (Lindert 2000; Piketty 2014). The share of the top 5% in pre-tax incomes was about 35% in 1750, roughly 40–42% in 1800 and 1867, nearly 38% circa 1900, falling to the 15–20% range circa 1950. The lopsided shares of the very top households coincided with sharp disparity of incomes across the classes. In 1867, for instance, the top decile received nearly 50% of national income, and the top centile nearly 29%; whereas, the share of the bottom 40% of households was about 17% and that of the lowest decile 3%. Even among franchised households, the share of the top 5% was about 52%, compared to only 11% for the bottom 40% (Lindert 2000).

The long-term trends in the shares of the top income earners and wealth holders were akin to the arc traced by the share of capital incomes in Fig. 6.2. The broader picture that surfaces from the trends in wealth and income distribution is: children in the earlier stages of the Transition grew up in a milieu of vast (and widening) income and wealth inequality, and a very steep socio-economic gradient. The ones brought up in the mid-transition stages in the early twentieth century received a situation in which the inequalities largely trended downward; the post-transition children grew up when the inequalities were least compared to other periods. The trends in inequality have been the subject of increasing interest over the last two decades, and scholars have begun studying how they can create, and can be created by, a polity that gives lopsided political power to the elite, an issue discussed in Sect. 6.5.

6.4 Together with Large Physiological Gaps

As the disparity of income and wealth grew during early industrialization, vast gaps in health across the social classes opened up. Based on the records of British peerage examined by T. H. Hollingsworth, Harris (2004b) reports that period-life expectancy at birth for the aristocracy circa 1750 was similar to that of the rest of the population (about 35–40 years). By 1850, the one for the peerage had pulled ahead to about 60 years, but the expectancy for the general population stayed at 35–40 years.

One source of the growing gap, of the gradient becoming steeper, was that the elite were able to adopt newer preventive methods such as inoculation before others (Deaton 2013). However, a cursory look at the nineteenth century disease classifications clarifies that inoculation could not have accounted for so large a change, as there were far too many diseases, even infectious diseases, and only against smallpox could there have been inoculation then (see Appendix). It is likelier that the growing affluence of the elite allowed them more protection from the severe epidemiologic situation the masses were facing. The changing gradient of health has been the cornerstone of a vast body of research on issues such as equity and the allocation of resources to health in the post-World War II era (Wilkinson 1986, 2001; Marmot and Wilkinson 1999; Marmot et al. 1991).

The book, however, focuses on a sliver of that broader issue, but one that is most pertinent to children (Case et al. 2002). The relevant issue here is not the population-wide gradient, per se, but how it had manifested in children's development. The goal is not to assess how the current health-related outcomes relate to current income and wealth distributions, or vice-versa. Instead, it is to assess if the current gradient, manifesting in children's development, mattered later on to the age-profiles over 50 years of the adult life cycle.

A childhood variable that can help assess such an issue needs to have two vital traits. It should reflect (1) how the inequality affected *only* childhood; and (2) how the inequality from then was channeled to the profiles over the adult years.

Average stature is a rare variable that possesses both. It is shaped only by what goes on in the childhood years. The influences from then get isolated, or packaged, so to speak, as stature neither shrinks nor grows after the onset of the adult years, and is carried forward into the adult-age profiles. Broader measures of inequality do not possess those traits of portability, making stature the more incisive instrument for the matter at hand.

In the nineteenth century, height was widely seen as a marker of wealth and status. So commonplace was the height gap across the classes that it found a place in one telling verse of Mrs. C. F. Alexander's children's hymn 'All things bright and beautiful' published mid-nineteenth century (Floud et al. 1990):

> The rich man in his castle,
> the poor man at the gate,
> God made them high and lowly
> and ordered their estate.

Over the past few decades, research in anthropometric history has been able to quantify the gaps between the 'high and lowly' by studying growth patterns of children from various socio-economic backgrounds recruited for military training and other services. For instance, Fig. 6.4 shows that 15-year-olds recruited by the Marine Society of London were much shorter than the boys of the same age

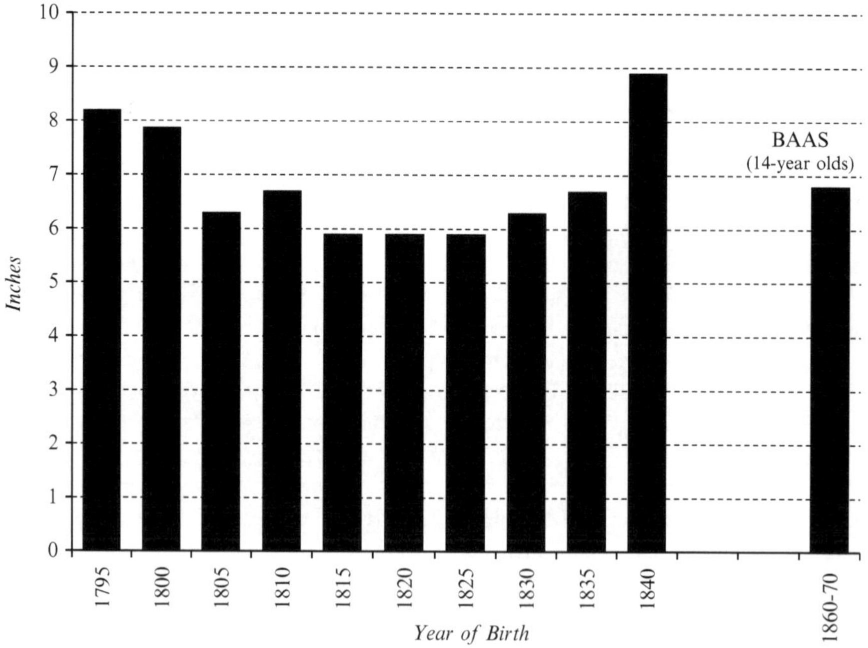

Fig. 6.4 The gap between the average heights of 15-year Old Sandhurst Recruits and Marine Society Recruits (Sources: Komlos (2008), Floud et al. (1990))

recruited by the Royal Military Academy at Sandhurst (Floud et al. 1990). The Marine Society boys were from the poorer social strata; the Sandhurst boys from the elite upper echelons. In the late eighteenth century, the Sandhurst boys were as much as 8.2 in. taller than the Marine boys. The gap between them shrank to about 6 in. in the first two decades of the nineteenth century, but widened again to 9 in. mid-century. Among 16-year olds in the 1840s, the Sandhurst boys were nearly 11 in. taller than the boys from the Marine Society. Height gaps at other ages in childhood show similar patterns, and in general boys from wealthier backgrounds grew sooner, taller, and with more velocity.

The elite were not just taller than the rest, they were on average nearly as tall as the recent modern averages. After adjusting for truncation of the height distributions, Komlos (2008) reports that the Sandhurst recruits were exceptionally tall for the time: at age 20, they were only 1.2 in. short of the average British male in the modern height distributions. By using the "fee" indicator for the richest of the rich, the "high-fee" segment averaged just six-tenths of an inch shorter than their modern average. Comparing them to the elite from other countries in the early nineteenth century, Komlos (2008) infers that the ones from Britain were the tallest and "clearly capable of overcoming the disadvantages brought about by a higher level of urbanization, higher population density, a more virulent disease environment, ..."

If the elite children from then grew up on average to be as tall as people have in recent decades, then it is possible that the long-term trends in the *average* height of entire birth cohorts may have been swayed less by the changes in the heights of the elite and more by the heights of children from other social classes. The average height of any cohort in that sense could be viewed as a proxy for how the disparity across the social classes tells on the disparity in the physiological growths of children.

Meanwhile, toward the end of the nineteenth century, in 1883, the Anthropometric Committee of the British Association of the Advancement of Science (BAAS), under Charles Roberts and Sir Francis Galton, reported that working-class 14-year old boys had remained about 5–6 in. shorter than the ones from the upper classes (though the sample for the working-class was small and the issue may not have been assessed accurately; Floud et al. 1990). By this time, officials seemed a lot more concerned than before about the state of the working classes. During the Boer War in 1899, for example, the "Inspector-General of Recruiting reported that up to 40% of those who were willing to join the army were unfit to serve..." (Harris, 2004a, p. 156). Recall from Chaps. 3 and 4, the adults during this time were the ones who grew up in the 1830s, 1840s, 1850s, and their poorer age-profiles were lifting the aggregate rate of non-communicable diseases.

Differences in heights across social classes provide "a powerful index of social and economic disadvantage" [in childhood] (Harris and Floud 1997, p. 118). Over the first half of the twentieth century, Harris (1989, 1994) examined records of individual school medical officers to assemble data on children's heights in more than 50 areas to find that average heights in different parts of Britain grew over

1908–1950, though there were still gaps between children from different social backgrounds. The growing average height over the period, nonetheless, signaled that children from the non-elite classes were likely narrowing the height gap. Moving further along to the 1950s and 1960s, among the 16–19-year olds born in the 1960s the gap between the uppermost and the lowermost classes had shrunk to about 1.2 in. (Knight and Eldridge 1984)—a sea change from the 6-to-11 in. gaps usual in the nineteenth century (also see Goldstein 1971; Rona et al. 1978; Gregory et al. 1990).

Average height has been known to be sensitive to the gini coefficient of income inequality in cross-country settings (Steckel 1995). However, within a country, over time, the relation between them may not be unidirectional, going from the gini to heights. If physiological development tells on people's lifetime earnings capacity—perhaps the lifetime itself—then the current gini could be influenced by the height gaps that had developed in the past.

Moreover, since height is irreversible, temporary changes can have lasting effects, resulting in a hysteresis-like outcome in aging. Hysteresis in economics refers to an outcome in labor markets when temporary unemployment becomes a structural one. Detached from the labor market for too long, the skills of the unemployed can depreciate or a new technology can make their skills obsolete (e.g. Blanchard and Summers 1986). A hysteresis-like outcome cannot be ruled out in how stature relates to aging. For instance, in successive mid-nineteenth century cohorts, mean heights fell and the profiles shifted up for the worse over the next 50 years. That change was permanent, even though the drop in the heights was temporary as the ensuing cohorts grew up to be taller on average.

The possibility of a hysteresis-like aftermath raises the issue of the type of within-cohort change accompanying the change in average height. The interpretation this book attaches to average stature is: wider the gap in the heights of children from the different social classes *within* a cohort, steeper the socio-economic gradient in the cohort's childhood years, poorer the average height of the cohort. All else equal, a downward trend in average height means that the gradient is becoming steeper, and the gap across the classes is widening; an upward trend in average height reflects the reverse.

The within-cohort view, however, is being linked to an aggregated profile. It must be that the profile, too, has a within-cohort parallel. To simplify, say there were only two classes, elite and non-elite. Consider the mid-nineteenth century cohorts whose average heights fell, and the age-profiles shifted up for the worse. The premise is: (1) the cohorts' average heights shrank as heights of children from non-elite classes fell more than the ones of elite children; and (2) the aggregated profile shifted up and became flatter than before as the non-elite profiles shifted up more and became flatter than the elite profiles. In general, all else equal, non-elite children gain or lose more height and sway the average height of the cohort; and the non-elite profiles sway the aggregated profile of a cohort more than the elite profiles. Chapter 8 revisits the matter by relating this to α_i and β_i of the profiles.

6.5 And of Embedded Political Inequality

The trends in inequality of income, wealth, and physiological growth coincided with vast inequity in the political sphere. Over the past two decades, there has been growing interest in how a polity can be of the type where few, who have monopoly-like political powers, can steer a large segment of the growing national income toward themselves.

The topic, however, goes back at least to the early studies of modern economic growth and income distribution in the 1950s. In his Presidential address to the American Economic Association, for instance, Simon Kuznets (1955) seemed optimistic that economic growth of the U.S. in first half of the twentieth century had coincided with more equitable incomes than before. Unaware of the trends preceding that unusual era, for a long time since economists saw income inequality as an upshot of economic growth and technological change instead of the other way around. One set of arguments at the time posited that by channeling incomes to high-saving classes, growing inequality may foster more economic growth as the high-savers are the ones who invest the most (Kaldor 1956, 1961).

That version proved increasingly dissatisfying in the 1990s as scholars looked further back in history, and at many poor countries today. They pointed out that the very wealthy can weigh in politically, and can block institutions such as mass public schooling, which is a source of human capital and economic growth. The wealth can be deployed politically either through laws, driven by patronage politics, or through regulatory capture that frames regulation tilted toward the interests of the elite. In this scenario, income inequality impedes economic growth (Engerman and Sokoloff 2005; Galor and Zeira 1993; Alesina and Rodrik 1994; Persson and Tabellini 1994; Bourguinon and Verdier 2000).

For instance, Galor et al. (2008) point out that landownership and wealth are usually quite lopsided in patriarchal societies of the type seen just before industrialization. Landowners stand to lose from labor being schooled too much, as the schooled may be lured away by more lucrative prospects. In contrast, new industrialists prefer that labor migrate to industrial towns for the new technologies to be profitable. If labor is of more value to industrialists than to landowners, and if the skills of labor complement industrial technology more than the rural one, technological change of the type that fuelled the Industrial Revolution may alter the nature relation between the classes. The tension between the landed elite and the masses in agrarian societies morphs into the one between the landed elite and the industrialists (Galor and Moav 2004, 2006). However, this theoretical outcome rests on the premise that industrialists and landowners are distinct people; or the erstwhile landowners have no stake in the new industrial ventures; or the burgeoning industrialists do not aspire ultimately to the political stature of the incumbent landed-elite to eventually guide policy in their own favor.

Nonetheless, the following quote illustrates the opposition of the landed elite to mass schooling. It is by Davies Giddy who opposed the 1807 Parliament bill on tax-based parochial schools:

> [G]iving education to the laboring classes of the poor ... would ... be prejudicial to their morals and happiness; it would teach them to despise their lot in life, *instead of making them good servants in agriculture*, and other laborious employment to which their rank in society had destined them; instead of teaching them subordination, it would *render them fractious and refractory, as was evident in the manufacturing counties*; it would enable them to read seditious pamphlets, vicious books, and publications against Christianity; it would render them insolent to their superiors ....Beside,....it would go to burden the country with a most enormous expense, and to load the industrious orders of society with still heavier imposts. [Parliamentary Debates [Hansard], 1807, Vol. IX, pp. 798–799; Lindert (2004), Vol. I; italics mine]

Remarking on the quote, Lindert (2004, p. 100) states: "Three premises behind this classic statement were all correct. Yes, education would cause laborers to leave agriculture for better jobs. Yes, education was seditious, in the sense that it would raise public opposition to landed Tory supremacy. And yes, paying for mass education would mean more taxes..."

According to the new set of theories, class tensions play out in the political arena where voting power—hence, political inequality—is critical. Vital reforms may not appear until there is a shift in the political set up, and one side out-votes the other, or until the interests of all sides coalesce. In another twist on the theme, Acemoglu and Robinson (2000) posit that elite-led governments of the early industrial era, especially the one in UK, eventually sought to cut the risk of social unrest and revolution, and offered political reforms such as the franchise to signal that the government would be more committed to 'redistribution' than it had been (also see Acemoglu and Robinson 2001, 2008; Acemoglu et al. 2001).

Theories of unequal political power, either as a cause of wealth and income inequality, or as a consequence of it, speak to what Peter Lindert has called the Robin Hood paradox: "redistribution toward the poor tends to happen least in those times and polities where it would seem most justified by the usual goals of welfare policy" (Lindert 2004, pp. 22–25, 1998). A central theme in the aforementioned theories appears to be that Robin Hood does not show up when needed most, likely because people might not have an effective political voice to summon him.

In the arena of public health, why Robin Hood failed to appear in the grim early stages of the Transition may be linked to political inequality as well. Szreter (1997) points out that new industrial technologies and migration of the masses to urban areas had disrupted longstanding social relations and the lines of authority established over previous centuries. However, nothing coherent had come up to replace them for decades. That created an administrative and political paralysis in cities through at least the first half of the nineteenth century. It took decades for the new interest in the urban areas to coalesce and for the reformers to push through legislations that benefited the masses.

Exploring the significance of politics to public health, he points out that many towns in the late-eighteenth and early-nineteenth century already had a coherent anti-aristocratic and anti-rentier voice of urban reform; so the opposition toward the elite was nothing new to the nineteenth century. New to the time was how that voice would join, albeit very slowly, the new bourgeois liberal causes in the industrial towns. Once they came together, the ranks of urban employers, artisans and workers became more coherently opposed to the rural landed elite who used to frame the laws.

The rising wealth of the commercial bourgeoisie, along with the growing ranks of the working classes, became harder to dismiss. The elite then gradually conceded constitutional reforms. The Reform Act 1832 electorally franchised some of the commercial bourgeoisie, though leaving the vast majority of people without the power to vote. The 1835 Municipal Reform Act created local representative governments in towns, which at the time had an electorate of small-scale property owners, but ones that still did not necessarily represent the interests of the masses. Nonetheless, the two Acts raised the size of the electorate from about 10% of adult males to 20% of all adult males (Szreter 2005; Cook and Stevenson 1983, 1996).

However, the Acts also fractured the burgeoning coalition that was until then opposed to the landed elite. The reforms produced an electorate more or less drawn from the class of small property owners; and the petty capitalist class, with their concern for 'economy,' became the new obstacle to municipal reforms as they favored commercially-viable ideas. Most public health infrastructure investments usually do not fall into that category.

In cities, the politics of water supply—a vital resource that underpins health—appears to have been a case in point. There had been growth in private water companies supplying water to businesses and wealthier homes in the first half of the nineteenth century. The technology for water supply to urban areas and for arterial sewage was also well understood as early as the 1840s in Edwin Chadwick's sanitary idea that tried to resolve the classic economic problem of the public good; of incentives and information costs; and of voter apathy (when costs of participation exceeded the diffuse benefits); of lowering the social costs stemming from the lack of hygiene; and of conditions under which to contract out sanitation to private participants or to open it up to franchise bidding; among other imperfections of the market in this arena (Chadwick 1842, 1859, 1885; Flinn 1965; Ekelund and Price 2012).

The blueprint of Chadwick's sanitary idea had homes connected to clean water supply as well as to a water-driven sewerage system. However, in the new municipal set up after the 1835 Municipal Reform Act, water supply had been connected, but the water-driven sewerage system had not. As late as 1871 the Royal Sanitary Commission found that most cities were just starting the sewerage part of the plan, despite the fact that there had been two major cholera outbreaks, and E. C. Snow in 1854 had shown that contaminated water had transmitted the epidemic (see Figs. 2.2 and 2.3, Chap. 2). The matter, nonetheless, remained unaddressed as the technology for treating the sewage flowing into the rivers was either not understood or not deployed because it was costly if internalized by private companies (Rosenthal 2014). The municipalized water after the 1848 Public Health Act, it appears, had been pursued more for commercial purpose than for public health goals. Further, the 1848 Act had opened up subsidized loans from the central government for water and sewerage projects. However, the local government, in the grip of the petty commercial class, rarely invoked the clause, and eventually had it revoked in an 1858 Act (Hassan 1985).

This type of dysfunction of municipal government has more twists than the one framed in some of the new theories of inequality and class, where the interests of industrialists and working classes naturally overlap in opposition to the interests of the landed elite. The political institutions of the time seem to have been beset by patronage politics connected to commercial houses. The local monopolies that it may have fostered may not have worked to the benefit of the urban masses. Even as early as 1822, a telling speech to the House of Commons, by Lord Russell, one of the leading voices for parliamentary reform, stated:

> I could, if I did not fear to fatigue the attention of the House, and if the thing were not so well known, read a number of letters clearly showing in many instances in which return of members to this House were returned by money only; by bribery the most direct...... One of the worst consequences of this system is the possession of power without responsibility. In fact, the individuals thus buying himself in, represents only the commercial house to which he belongs. I remember on one occasion, a member who got into the House by dint of money... came to me and assured me that he had no wish whatsoever to enter parliament, but that he did so to oblige his partners in trade. Now, that is exactly the kind of representative that I do not wish to see in the House. I do not wish to see men returned here for commercial houses, representing only their partners, and naturally anxious to oblige the government in order to procure patronage and favor for their establishment. (Hansard 1822 pp. 62–64; Lizzeri and Persico 2004)

In the decades leading up to the mid-nineteenth century, the voices of the reformers seems to have made more difference to the passage of legislation beneficial for the masses. Akin to the modern-day 'technocrats,' leading public health reformers of the time (Chadwick, Duncan, Farr and Simon) seemed to have a practical view that the situation could be improved by the application of science and by reform, but would encounter political hurdles. Built around the 'civic gospel' the period also saw a growing new willingness to combat hazards to health. This often benefitted from effective leadership by some urban propertied elite, who undercut the 'economy' argument by pointing out that failure to reform was a false economy as it took a hefty toll on worker efficiency (Fraser 1976, 1993; Harris 2004a). One example of the growing new persuasion was the Sanitary Act 1866 that expanded the legal definition of 'nuisances' to health and boosted the power of the central government while increasing the obligations of the local ones.

The Reform Act 1867 and the Municipal Franchise Act 1869 then finally appear to have undercut the petty bourgeoisie's grip on municipal government. The size of the electorate jumped from about 700,000, (or just 20% of adult men) in 1832, to about two million (or to roughly 33% of all adult males), likely tipping the scales in the favor of reform at the central and the local levels (Szreter 1997, 2005).

For the various stages of the Transition, Fig. 6.5 provides an aggregate-level view of the near-paralysis of reform in the first half of the nineteenth century, and of the ascendancy of reform in the second half. The step-like line is the democracy score calculated by the Polity IV database. Running from 1 through 10, the score 7 to 10 signifies full democracy, defined as one in which "political participation is unrestricted, open and fully competitive; executive recruitment is elective; and constraints on the chief executive are substantial."

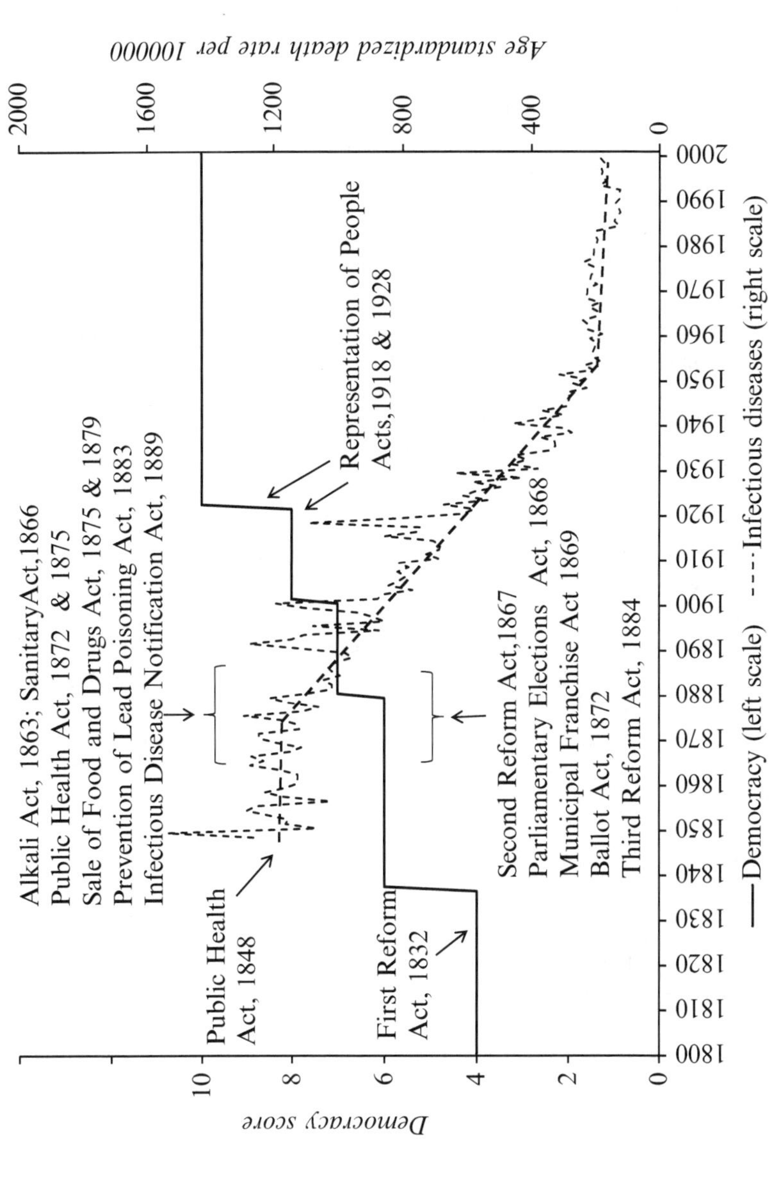

Fig. 6.5 Democracy score, political and public health reforms, and infectious diseases during the transition (Sources: Polity IV database, April 2013, www.systemicpeace.org; HMSO; author's calculations)

The score is based on issues such as how competitively the executive is recruited (e.g. election, hereditary selection, or dual executives where one is elected while the other selected); how open is the recruitment; if the politically active can in principle attain the executive position; how constrained is the chief executive; and how competitive is political participation. To avoid clutter, Fig. 6.5 states only few of the principal electoral changes since the nineteenth century (for others, see Craig 1976, p. 173).

Reduction in political inequality, or the dilution of the elites' political power, preceded each jump in the democracy score. The first one was after The Reform Act 1832 that expanded the franchise modestly (from 10% of adult men to 20%), but left out the majority. The Second Reform Act 1867 boosted the electorate from 20% of adult males to 33%, the main source of the second jump, though women still could not vote. The Ballot Act 1872 started the secret ballot. The Third Reform Act 1884 set up more uniformity than before, allowing only one Member of Parliament (MP) from most districts, increasing the representation of the urban masses; previously some small rural areas had two MPs while some cities had none. It coincided with an increase in the electorate from 33% of all adult males in 1867 to about 66% of adult males in 1885. The Representation of People Act 1918 ended property qualifications for voting, and gave women of age 30 years or older the right to vote; the electorate by this time was 100% of all adult males 21 years or older, and 50% of adult females age 21 years or older. Equal franchise for all adults was achieved with the Representation of People Act 1928, finally taking the democracy score to 10.

Figure 6.5 conveys that children that grew up in the early stages of the transition received a political milieu in which the masses were disallowed the vote. Children from the transition stage grew up in a political setting where the power of the elite was increasingly being diluted as the franchise was gradually being extended. Children in the post-transition stage grew up under equal adult franchise.

6.6 Its Reform Reveals a Natural Experiment

Alongside the democracy score in Fig. 6.5 is the long-term trend in infectious diseases since circa 1850. It is striking how closely the decrease in political inequality since The Second Reform Act 1867, and the Municipal Franchise Act 1869, coincided with a cluster of major public health reforms that may have swayed the broader direction of infectious diseases.

The clusters of political and public health reforms strongly indicate that partial removal of one type of inequality (of voting power) may reduce another type of inequality (sewerage, clean water, epidemiologic situation in general (Sen 1984, 1999; Deaton 2013)). This coinciding pattern exemplifies 'co-realizable capabilities' discussed in Chap. 3, Sect. 3.5: the capability to vote may have helped realize the capability to improve the epidemiologic situation, and may have helped boost the development of children brought up since.

It helps to briefly outline some of the reforms. The Alkali Act of 1863 was passed to cut atmospheric pollution and emissions from Alkali factories; legislation in 1876 barred the dumping of industrial waste and refuse in local rivers. The Factory Act of 1864 regulated ventilation of factories; the Factory and Workshop Act 1878 barred women and young children from being employed in lead factories; and the Prevention of Lead Poisoning Act of 1883 set up strict standards for lead workers.

The Public Health Act 1872 created for each area in England and Wales a sanitary authority with its own administration and its own statutory Medical Officer of Health. The Public Health Act 1875, by far the largest of all reforms, unified about two decades of legislation and became the foundation of public health policy until the advent of the National Health Service in the post-transition era (Frazer 1950, pp. 117–25). Apart from water supply, sewerage and cleanup of hazards (the 'miasmas' thought to be the source of disease), local authorities in the 1870s began special hospitals for infectious diseases (Wohl 1983). To rein in potential outbreaks, the Infectious Disease Notification Act 1889 required notification of a list of such diseases to authorities.

It seems very likely that the political reforms in the 1860s and 1870s had emboldened the politicians and the reformers to push through critical public health legislation in the interest of the masses who voted them into office. Lindert's Robin Hood had likely arrived. However, the type of 'redistribution' it may have fostered may not have been a one-shot issue where taxes levied from one segment of society are transferred to another for consumption. The resources raised were used for infrastructure investments that benefited the masses for long. They may also have fostered a beneficial long-term intertemporal link between children's development and the profiles of aging. Chapter 9 will return to this issue of political roots in the context of healthcare spending.

Meanwhile, the timing of the reforms offers a natural experiment for studying cross-cohort aging. The laws are situated roughly on the cusp of the regimes. Before the reforms was the Malthusian-like epidemiologic situation; after the reforms was the transition stage. The Malthusian cohorts of children were raised before the reforms; the transition ones after then. The next chapter studies cross-cohort aging by exploring this natural experiment. It uses the labels 'Malthusian' and 'pre-reform,' 'transition' and 'post-reform' interchangeably.

In sum, for studying the relation of children's development to aging, it is vital to recognize that the socio-economic and political conditions in which children were being raised were changing radically since the early nineteenth century. Successive cohorts were raised under increasing prosperity since the Industrial Revolution, although the volatility of per capita incomes in their childhoods had varied. They received inequalities of various forms. Inequality, in general, was likely widening for the childhoods in the Malthusian or the pre-reform stage, and likely narrowing for the childhoods in the transition or the post-reform stage. Inequality in income and wealth, especially in the ownership of capital, was likely behind the trends in the functional distribution of national incomes; the share of capital climbed sizably in the Malthusian stage before shrinking gradually over the transition stage, tracing a path similar to that of the income share going to the elite. Meanwhile, average heights, reflecting physiological inequalities, trended down during the mid-nineteenth century in the Malthusian stage, to improve near the decades of reforms and thereafter. Political inequality, evident in the unequal voting power for the masses reduced sizably since the Second Reform Act 1867 and the Municipal Franchise Act 1869, coinciding roughly with some of the more fundamental public health reforms in UK's modern history.

References

Acemoglu, D., & Robinson, J. (2000). Why did the West extend the Franchise? Democracy, inequality and growth in historical perspective. *The Quarterly Journal of Economics, 115*(4), 1167–1199.

Acemoglu, D., & Robinson, J. (2001). A theory of political transitions. *American Economic Review, 91*, 938–963.

Acemoglu, D., & Robinson, J. (2008). Persistence of power, elites and institutions. *American Economic Review, 98*(1), 267–293.

Acemoglu, D., Johnson, S., & Robinson, J. (2001). The colonial origins of comparative development: An empirical investigation. *American Economic Review, 91*, 1369–1401.

Aksan, A.-M., & Chakraborty, S. (2012). *Twin transitions*. Working paper, Economics Department, University of Oregon.

Alesina, A., & Rodrik, D. (1994). Distributive politics and economic growth. *The Quarterly Journal of Economics, 108*, 465–490.

Allen, R. (2009). Engels' pause: Technical change, capital accumulation, and inequality in the British industrial revolution. *Explorations in Economic History, 46*(4), 418–435.

Arora, S. (2001). Health, human productivity, and long-term economic growth. *Journal of Economic History, 61*(3), 699–749.

Atkinson, A. B., & Piketty, T. (2007). *Top incomes over the twentieth century: A contrast between continental-European and English-speaking countries*. Oxford: Oxford University Press.

Atkinson, A. B., & Piketty, T. (2010). *Top incomes: A global perspective*. Oxford: Oxford University Press.

Atkinson, A. B., Piketty, T., & Saez, E. (2011). Top incomes in the long-run of history. *Journal of Economic Literature, 49*(1), 3.

Blanchard, O., & Summers, L. H. (1986). Hysteresis and the European unemployment problem. In S. Fischer (Ed.), *NBER macroeconomics annual* (Vol. 1). MA: Cambridge.

Bourguinon, F., & Verdier, T. (2000). Oligarchy, democracy, inequality and growth. *Journal of Development Economics, 62*(2), 285–313.

Case, A., Lubotsky, D., & Paxon, C. (2002). Economic status and health in childhood: The origins of the gradient. *American Economic Review, 92*(5), 1308–1334.

Chadwick, E. (1842[1965]). *Report to Her Majesty's principal secretary of State for the Home Department, from the Poor Law Commissioners, on an Inquiry into the Sanitary Condition of the Laboring Population of Great Britain*. London: W. Clowes and Sons. Reprinted as *Report on the Sanitary Condition of the Laboring Population of Great Britain*, edited with an introduction by M. W. Flinn. Edinburgh: Edinburgh University Press.

Chadwick, E. (1859). Results of different principals of legislation and administration in Europe; Of competition for the field, as compared with competition within the field of service. *Journal of the Royal Statistical Society, 22*, 381–420.

Chadwick, E. (1885). *Of the evils of disunity of central and local administration especially with relation to the metropolis and also on the new centralization for the people together with improvements in codification and in legislative procedure*. London: Longmans, Green and Co.

Cook, C., & Stevenson, J. (1983). *Longman handbook of modern British history 1714–1980*. London: Longman.
Cook, C., & Stevenson, J. (1996). *Longman companion to Britain since 1945*. London: Longman.
Crafts, N. F. R. (1985a). *British economic growth during the industrial revolution*. Oxford: Clarendon Press.
Crafts, N. F. R. (1985b). English workers' living standards during the industrial revolution: Some remaining problems. *Journal of Economic History, 45*, 139–144.
Crafts, N. F. R. (1989). Real wages, inequality and economic growth in Britain 1750–1850: A review of recent research. In P. Scholliers (Ed.), *Real wages in nineteenth and twentieth century Europe: Historical and comparative perspectives*. Oxford: Berg.
Craig, F. W. S. (1976). *British electoral facts 1885–1975*. London: The McMillan Press.
Davis, J., & Tanner, D. (1969). The borough franchise after 1867. *Historical Research, 69*, 306–327.
Deaton, A. (2013). *The great escape*. Princeton: Princeton University Press.
Ekelund, R. B., Jr., & Price, E. O., III. (2012). *The economics of Edwin Chadwick: Incentives matter*. MA: Edward Elgar.
Engerman, S., & Sokoloff, K. (2005). The evolution of suffrage institutions in the Americas. *Journal of Economic History, 65*(4), 891–921.
Flinn, M. W. (1965). *Report of the sanitary conditions of the laboring population of Great Britain by Edwin Chadwick, 1842*. Edinburgh: Edinburgh University Press.
Floud, R., Wachter, K., & Gregory, A. (1990). *Height, health, and history: Nutritional status in the United Kingdom, 1750–1980*. Cambridge: Cambridge University Press.
Fraser, D. (1976). *Urban politics in Victorian England: The structure of politics in Victorian cities*. Leicester: Leicester University Press.
Fraser, H. (1993). Municipal socialism and social policy. In R. J. Morris & R. Rodger (Eds.), *The Victorian city: A reader in British urban history, 1820–1914* (pp. 258–280). London: Longman.
Frazer, W. M. (1950). *History of English public health 1834–1939*. London: Bailliere, Tindall and Cox.
Galor, O., & Moav, O. (2004). From physical to human capital accumulation: Inequality and the process of economic development. *Review of Economic Studies, 71*, 1001–1026.
Galor, O., & Moav, O. (2006). Das human kapital: A theory of the demise of the class structure. *Review of Economic Studies, 73*, 85–117.
Galor, O., & Zeira, J. (1993). Income distribution and macroeconomics. *Review of Economic Studies, 60*, 35–52.
Galor, O., Zeira, J., & Vollrath, D. (2008). Inequality in landownership, the emergence of human capital-promoting institutions, and the great divergence. *Review of Economic Studies, 75*, 143–179.
Goldstein, H. (1971). Factors affecting the height of seven-year old children: Results from the National Child Development Study. *Human Biology, 43*, 92–111.
Gregory, J., Foster, K., Tyler, H., & Wiseman, M. (1990). *The dietary and nutritional survey of British adults: A survey carried out by the Social Survey Division of OPCS with dietary and nutritional evaluations by the Ministry of Agriculture, Fisheries, and Food, and the Department of Health*. London: HMSO.
Hansard. www.parliament.uk\publications and records.
Harris, B. (1989). *Medical inspection and the nutrition of school children in Britain 1900–50*. Ph. D. thesis, University of London.
Harris, B. (1994). The height of school children in Britain 1900–50. In J. Komlos (Ed.), *Stature, living standards and economic development; Essays in anthropometric history* (pp. 25–38). Chicago: University of Chicago Press, New York.
Harris, B. (2004a). *The origins of the British welfare state: Social welfare in England and Wales 1800–1945*. Basingstoke: Palgrave, Macmillan.
Harris, B. (2004b). Public health, nutrition and the decline of mortality: The Mckeown thesis revisited. *Social History of Medicine, 17*, 379–407.

Harris, B., & Floud, R. (1997). Health, height, and welfare: Britain, 1700–1980. In R. H. Steckel & R. Floud (Eds.), *Health, welfare and industrialization*. Chicago: University of Chicago Press.
Hassan, J. A. (1985). The growth and impact of British water industry in the nineteenth century. *Economic History Review, 38*, 531–547.
Horrell, S., & Humphries, J. (1995). The exploitation of little children: Children's work and the family economy in the British industrial revolution. *Explorations in Economic History, 32*, 849–880.
Horrell, S., Humphries, J., Humphries, J., & Voth, H.-J. (2001). Destined for deprivation: Human capital formation and intergenerational poverty in nineteenth century England. *Explorations in Economic History, 38*, 339–365.
Howitt, P. (2005). Health, human capital and economic growth: A Schumpeterian perspective. In G. Lopez-Casasnovas, B. Rivera, & L. Currias (Eds.), *Health and economic growth: Findings and policy implications* (pp. 17–40). Cambridge: MIT press.
Humphries, J. (2013). *Childhood and child labor in the British industrial revolution*. New York: Cambridge University Press.
Kaldor, N. (1956). Alternative theories of distribution. *Review of Economic Studies, 23*, 94–100.
Kaldor, N. (1961). Capital accumulation and economic growth. In F. A. Lutz & D. C. Hague (Eds.), *The theory of capital*. New York: St. Martin Press.
Knight, I., & Eldridge, J. (1984). *The heights and weights of adults in Great Britain: Report on a survey carried out on behalf of the Department of Health and Social Security among adults aged 16–64*. London: HMSO.
Komlos, J. (2008). On English pygmies and giants: The physical stature of English youth in the late 18th century and early 19th centuries. *Research in Economic History, 25*(2), 149–168.
Kuznets, S. (1955). Economic growth and income inequality. *American Economic Review, 45*(1), 65–94.
Lindert, P. (2000). Three centuries of inequality in Britain and America. In A. B. Atkinson & F. Bourguinon (Eds.), *Handbook of income distribution* (Vol. 1, pp. 176–316). Amsterdam: Elsevier Science.
Lindert, P. (2004). *Social spending and economic growth since the eighteenth century, Volume 1 and 2*. New York: Cambridge University Press.
Lindert, P., & Williamson, J. (1983). English workers' living standards during the industrial revolution: A new look. *Economic History Review, 36*, 1–25.
Lindert, P., & Williamson, J. (1985a). English workers' living standards during the industrial revolution: A new look. In J. Mokyr (Ed.), *The economics of the industrial revolution* (pp. 177–205). Totowa: Rowan and Allanheld.
Lindert, P., & Williamson, J. (1985b). English workers' real wages: A reply to Crafts. *Journal of Economic History, 45*, 145–153.
Lizzeri, A., & Persico, N. (2004). Why did the elites extend the Suffrage? Democracy and the Scope of Government, with and application to Britain's 'age of reform'. *The Quarterly Journal of Economics, 119*(2), 707–765.
Maddison, A. (2001). *The world economy: A millennial perspective*. Paris: OECD.
Marmot, M., & Wilkinson, R. G. (Eds.). (1999). *The social determinants of health*. Oxford: Oxford University Press.
Marmot, M., Davey-Smith, G., Stansfield, S., et al. (1991). Health inequalities among British civil servants: The Whitehall II Study. *Lancet, 337*, 1387–1393.
McCloskey, D. N. (1970). Did Victorian Britain fail? *Economic History Review, 23*(3), 446–459.
Mitchell, B. R. (2011). *British historical statistics*. Cambridge: Cambridge University Press.
Persson, T., & Tabellini, G. (1994). Is inequality harmful for growth? *American Economic Review, 84*, 600–621.
Piketty, T. (2014). *Capital in the twenty-first century*. Cambridge, MA: The Belknap Press of Harvard University Press.
Rona, R. J., Swan, A. V., & Altman, D. G. (1978). Social factors and height of primary school children in England and Scotland. *Journal of Epidemiology and Community Health, 32*, 147–154.

Rosenthal, L. (2014). *The river pollution dilemma in Victorian England: Nuisance law versus economic efficiency*. Farnham: Ashgate Publishing.
Sen, A. K. (1984). The living standard. *Oxford Economic Papers, 36*, 74.
Sen, A. K. (1999). *Commodities and capabilities*. Oxford: Oxford University Press.
Steckel, R. H. (1995). Stature and the standard of living. *Journal of Economic Literature, 33*(4), 1903–1940.
Szreter, S. (1997). Economic growth, disruption, deprivation, disease, and death: On the importance of politics of public health for development. *Population and Development Review, 23*(4), 693–728.
Szreter, S. (2005). *Health and wealth: Studies in history and policy*. Rochester: University of Rochester Press.
Tunzelmann, N. V. (1979). Trends in real wages, 1750–1850, revisited. *Economic History Review, 32*, 33–49.
Wilkinson, R. G. (Ed.). (1986). *Class and health*. London: McMillan. 1976.
Wilkinson, R. G. (Ed.). (2001). *Mind the gap: Hierarchies, health, and human evolution*. New Haven: Yale University Press.
Williamson, J. G. (1990). *Coping with city growth during the Industrial Revolution*. Cambridge: Cambridge University Press.
Wohl, A. S. (1983). *Endangered lives: Public health in Victorian Britain*. Cambridge, MA: Harvard University Press.

Chapter 7
In the Natural Experiment

7.1 The Selection Bias Likely Dissipates

The outcome of interest is the aging potential across cohorts. Denoted by β, it is measured by the slope of the age-profiles of non-communicable diseases over age 25–74. All else constant, larger β stands for better potential. Meanwhile, the natural experiment is the cluster of political and public health reforms in the late 1860s and in the 1870s. Denote 'reform' by the binary indicator r_i that stays 0 for the period before the reforms or 'pre-reform,' and switches to 1 for the period after the reforms or 'post-reform.' The birth decades D_i are the cohorts' fixed and independent traits.

Say, for the moment, there was just one pre-reform Malthusian and one post-reform transition cohort. An ideal assessment of any natural experiment requires two potential outcomes for each cohort. One of the outcomes would be based on reforms in the childhood years, and the other would be based on no reforms in the childhood years (Heckman 2008; Angirst and Pischke 2009; Wooldridge 2009). For any cohort i they are expressed as:

$$profile\ slope = \begin{cases} \beta_{0i} & if \quad r_i = 0 \\ \beta_{1i} & if \quad r_i = 1 \end{cases} \tag{7.1}$$

Here, β_{0i} determines cohort i's aging potential if its childhood years were not exposed to reform ($r_i = 0$), and β_{1i} is the potential—of that same cohort—if its childhood years were exposed to reform ($r_i = 1$). The difference between the two βs

$$\beta_{1i} - \beta_{0i} \tag{7.2}$$

is said to be the reform's influence on cohort i. However, the aging potential of the Malthusian cohort if its childhood had been exposed to reform cannot be known; so the first element (β_{1i}) is missing for this cohort. Similarly, for the transition cohort,

S. Arora, *The Transitions of Aging*, International Perspectives on Aging 12,
DOI 10.1007/978-3-319-14403-0_7

only the aging potential post-reform is known and how the potential might have fared without reform in the childhood years (β_{0i}) cannot be seen.

Instead of one pre-reform and post-reform cohort, if there were several of them in each era, the ideal measure is an average or the expected value of the gaps between each cohort's two βs, given reform, or

$$E[\beta_{1i} - \beta_{0i}|r = 1] \tag{7.3}$$

which is even a taller order for history to fulfill, since only one outcome per cohort came to pass. With the 'ideal' data missing, it then becomes tempting to compare the expected values only of the observed set of βs. That is, average the actual post-reform cohorts' βs, and average the actual pre-reform cohorts', calculate the difference between them, thus

$$E[\beta_i|r_i = 1] - E[\beta_i|r_i = 0] \tag{7.4}$$

and find out if the reforms had swayed the outcome. The ideal way, however, warns that by giving-in to that temptation one lets a bias creep into the assessment, as expression 7.4 is actually made up of two components

$$\underbrace{E[\beta_{1i} - \beta_{0i}|r_i = 1]}_{\text{I}} + \underbrace{\{E[\beta_{0i}|r_i = 1] - E[\beta_{0i}|r_i = 0]\}}_{\text{II}}$$

The first one is the true gauge of the reform's influence. The second one is a selection bias that would cloud the assessment. The bias is the average βs of the cohorts had their childhoods been exposed to reform, minus, the average βs of those same cohorts if their childhoods had not been exposed to reforms.

The bias acknowledges the possibility that cohorts exposed to reform might have done well, perhaps even just as well, had there been no reforms in their childhood years at all. It also means that the reforms could have been a mere coincidence. The cohorts faring better since the reforms were somehow capable of doing better anyway. Without acknowledging this possibility, too much credit can be given to the reforms.

Such bias is of major concern when benefits of a policy or a program have to be assessed. To see it better, say a program had been set up to foster a trait. Say also it was set up on the premise that people who do not have trait might come to acquire it through the program, and then benefit. However, some people who benefit from the program may have already possessed the trait and ended up doing well $[\beta_{0i}|r_i = 1]$, but since they had the trait on their own, they might have done well without a nudge from the program at all $[\beta_{0i}|r_i = 0]$. If so, some of what would appear as the program's influence could actually have been a mere selection bias (II)—people who already had the trait, but ended up doing well having enrolled in the program. This is roughly akin to saying that a student's performance is not entirely attributable to the great teacher, but also to more capable students possibly having enrolled in the teacher's class. The role of student capability, the source of the selection bias, needs to be isolated to see the teacher's true influence (term I).

For the matter at hand, however, such a bias can occur only if the cohorts have the ability to select their birth decades. It could arise, for instance, if the Malthusian cohorts might have fared better under the reform era $[\beta_{0i}|r_i = 1]$, but that could only happen if they had the ability to change the birth decade to the era. Likewise, the possibility that the transition cohorts might have done better without reform $[\beta_{0i}|r_i = 0]$ hinges on their ability to switch the birth decade to the pre-reform era. This type of selection seems very unlikely, if possible at all.

Expressed technically, the cohorts' birth decades D_i and the occurrence of reforms r_i are mutually independent. Cohorts were born when they were born, and not because the reforms were offered at a specific time. Likewise, the reforms were passed when they were passed and not deliberately timed to leave the Malthusian cohorts out and to include the rest. Conditioned on the mutual independence, the selection bias likely dissipates. If so, expression 7.4 and its first component (I) would be expressed as:

$$E[\beta_i|D_i, r_i = 1] - E[\beta_i|D_i, r_i = 0] = E[\beta_{1i} - \beta_{0i}|D_i] \tag{7.5}$$

Or, the average of the transition cohorts' βs minus the average of the Malthusian cohorts' βs—that initial temptation—equate to the right-hand side, the reform's true effect, given reform (r_i) occurred independently of the cohorts' childhood years (D_i).

Further, the political reforms were neither intended nor passed to directly sway the outcome of interest: the non-communicable diseases in the age-profiles decades down the road. The pressing issue the Second Reform Act or the Municipal Franchise Act tackled was the disparity in political voice and in the right to vote. The urgent issue addressed by the major Public Health Acts was the disparity in access to water, sewerage, and to rein in miasmatic diseases by purging the miasmas. Moreover, the profiles of non-communicable diseases were actually shifting up during the second half of the nineteenth century. The growing aggregate period-rates seemed unfazed by the reforms, possibly because the issues the reforms addressed were distant from what had caused the rates of such diseases to trend up.

The reforms, however, may have influenced the outcome indirectly. They could have led to less disparity in children's physiological developments (e.g. better mean heights), resulted in fewer infections in children, or they may have cut the extreme volatility of such diseases prevalent then (e.g. containment of outbreaks). Growing up in the newer environment, successive cohorts of children may have aged along new and better age-profiles. This indirect channel is an issue of how the influence of reforms may have been transmitted inadvertently to the new profiles.

Instead of being a shortfall, however, the indirect influence, or the inadvertent transmission, is just what makes the experiment worth exploiting.

In sum, the reforms are a compelling natural-experiment-type of situation as they were (1) sufficiently exogenous and not intended to directly sway the outcome

Table 7.1 Pre-reform and post-reform cohorts by birth decade

	Regime	Set I	Set II	Set III	Set IV
Pre-reform	Malthusian		1795–1804		1796–1805
$r_i = 0$	Malthusian	1800–1809	1805–1814	1801–1810	1806–1815
$N_i = 26$	Malthusian	1810–1819	1815–1824	1811–1820	1816–1825
	Malthusian	1820–1829	1825–1834	1821–1830	1826–1835
	Malthusian	1830–1839	1835–1844	1831–1840	1836–1845
	Malthusian	1840–1849	1845–1854	1841–1850	1846–1855
	Malthusian	1850–1859	1855–1864	1851–1860	1856–1865
Post-reform	Early-transition	1860–1869	1865–1874	1861–1870	1866–1875
$r_i = 1$	Early-transition	1870–1879	1875–1884	1871–1880	1876–1885
$N_i = 28$	Early-transition	1880–1889	1885–1894	1881–1890	1886–1895
	Mid-transition	1890–1899	1895–1904	1891–1900	1896–1905
	Mid-transition	1900–1909	1905–1914	1901–1910	1906–1915
	Late-transition	1910–1919	1915–1924	1911–1920	1916–1925
	Late-transition	1920–1929	1925–1934	1921–1930	1926–1935
$r_i = 1$	Post-transition	1930–1939	1935–1944	1931–1940	1936–1945
$N_i = 8$	Post-transition	1940–1949	1945–1954	1941–1950	1946–1955

of interest (non-communicable disease decades later); and (2) the cohorts' childhood years and the reforms were mutually independent: there could neither have been a realistic way of self-selection and the bias it entails; nor is it very likely that the laws were framed and timed to deliberately favor children from a set of cohorts but not others. The set up thus allows the use of Eq. 7.5 for an initial assessment.

Table 7.1 shows how the birth decades fall into the pre-reform and the post-reform eras. All transition stage cohorts are post-reform, with the 1860s and 1870s cohorts on the cusp. Moreover, the reforms had set into motion changes that cumulated over the long term. Since it is unrealistic to expect the entire nation's infrastructure to have been built at once, the progress over the decades likely embraced a growing fraction of children. Consequently, children that grew up in the later stages of the transition likely received a better epidemiologic setting than the ones brought up in the earlier stages. So the edge, say, the late-transition cohorts might have had over the Malthusian cohorts could have been at least as large if not more than the edge the mid-transition ones had over the Malthusian cohorts.

Further, in the post-reform era, Table 7.1 sets apart the post-transition cohorts, as children from the 1930s and 1940s received two more regime-defining changes that can be viewed in a similar natural experiment-type set up. They received (1) antibiotics and vaccines, which were new to the world of medicine then; and (2) the National Health Service, which was available during the childhood years of at least some, though not all, who were born in the 1930s and 1940s. The distinction between the post-transition cohorts and the Malthusian ones thus involves (1) post-reform versus pre-reform; (2) universal health insurance versus at best imperfect health insurance for the Malthusian cohorts; (3) new antibiotics and new vaccines for infectious diseases versus almost none available to the Malthusian cohorts (except smallpox inoculations).

7.2 And the Reforms May Have Boosted Aging Potentials

Tables 7.2, 7.3, 7.4, 7.5, 7.6, and 7.7 show the results. The information in the top segment of Table 7.2 is based on $\hat{\beta}_i$ and $\hat{\alpha}_i$ in all four sets presented earlier in Tables 3.9, 3.10, 3.11, and 3.12. The second column, $E[\hat{\beta}_i r_i]$, shows the average, $\overline{\hat{\beta}}_i$, of each group mentioned in the first column. The subscript $i = 0$ and the regime marker $r_i = 0$ denote pre-reform. The third column has the variances of the $\hat{\beta}_i$ in each group, calculated as

$$\hat{\sigma}_i^2 = \frac{1}{N_i - 1}\sum\nolimits_i \left(\hat{\beta}_i - \overline{\hat{\beta}}_i\right)^2 \tag{7.6}$$

The fourth column has the variances divided by the number of cohorts in the group. The information in the top-left segment then feeds the calculations shown in the bottom-left segment that formally tests if the differences were statistically significant. The t-statistic is

$$\frac{t = \overline{\hat{\beta}}_i - \overline{\hat{\beta}}_0}{\Phi} \tag{7.7}$$

where the first term in the numerator is the average $\hat{\beta}_i$ of the said post-reform cohorts; the second one is the average $\hat{\beta}_0$ of the pre-reform cohorts; and the denominator is

$$\Phi = \sqrt{\frac{\hat{\sigma}_i^2}{N_i} + \frac{\hat{\sigma}_0^2}{N_0}} \tag{7.8}$$

with $\hat{\sigma}_i^2$ as the variance of $\hat{\beta}_i$, $\hat{\sigma}_0^2$ the variance of the $\hat{\beta}_0$, and N the number of cohorts in each group. The null hypothesis (no effect of reforms) and the alternative hypothesis are

$$H_0 : \overline{\hat{\beta}}_i - \overline{\hat{\beta}}_0 = 0$$
$$H_1 : \overline{\hat{\beta}}_i - \overline{\hat{\beta}}_0 \neq 0$$

The top-right segment of Table 7.2 goes through the same exercise for $\hat{\alpha}_i$. The null and alternative hypotheses are

$$H_0 : \overline{\hat{\alpha}}_i - \overline{\hat{\alpha}}_0 = 0$$
$$H_1 : \overline{\hat{\alpha}}_i - \overline{\hat{\alpha}}_0 \neq 0$$

and the rest of the terms mean the same things as the left-side of the table, only calculated for the $\hat{\alpha}_i$ instead.

The data reject the null hypotheses for both parameters. The results strengthen as one goes further into the transition stage. They support the premise that, all else constant, children from the later stages of the transition had a greater edge over the pre-reform cohorts than the children from the earlier stages, implying that the cumulating change mattered, and that the influence was not a one-shot outcome affecting only the cohorts immediately after the reforms.

Further, the information in the Table 7.2 can be used for calculating the results for, say, the post-transition versus the mid-transition cohorts: the gap in their $\overline{\hat{\beta}_i}$ is 0.0072, the Φ is the square root of (0.00000103 + 0.00000070) or 0.001315; and the resulting t-statistic, 0.0072 divided by 0.001315, is 5.47.

Using the same format, Tables 7.3, 7.4, 7.5, 7.6, and 7.7 show results for some of the sub-categories of non-communicable disease. The case for circulatory system diseases in Table 7.3 is similar to that of the aggregate. Meanwhile, the exception flagged in Chap. 3 for the diseases of the nervous system appears here in the results shown in Table 7.4. Consider the results for its $\hat{\alpha}_i$. The initial state shifted down substantially, suggesting a much stronger start for the newer cohorts; and, the data assesses the shift to be statistically significant. However, the bottom-left side of the Table says that the slope for the mid-transition cohorts had on average become flatter than the one of the pre-reform cohorts. Instead of being a loss in aging

Table 7.2 Pre-reform vs. post-reform $\hat{\beta}_i$ and $\hat{\alpha}_i$, aggregate non-communicable diseases

	$E[\hat{\beta}_i \mid r_i]$	$\hat{\sigma}_i^2$	$\hat{\sigma}_i^2/N_i$	$E[\hat{\alpha}_i \mid r_i]$	$\hat{\sigma}_i^2$	$\hat{\sigma}_i^2/N_i$
Pre-reform cohorts, $r_i = 0$	0.0741	0.0000218	0.00000084	24.7	58.1	2.2
Post-reform cohorts, $r_i = 1$	0.0784	0.0000234	0.00000083	14.5	36.2	1.3
Mid-transition cohorts, $r_i = 1$	0.0787	0.0000056	0.00000070	12.5	3.5	0.43
Late-transition cohorts, $r_i = 1$	0.0842	0.0000088	0.00000110	7.7	3.0	0.37
Post-transition cohorts, $r_i = 1$	0.0859	0.0000082	0.00000103	5.3	0.85	0.11
	$\overline{\hat{\beta}_i} - \overline{\hat{\beta}_0}$	Φ	t-statistic	$\overline{\hat{\alpha}_i} - \overline{\hat{\alpha}_0}$	Φ	t-statistic
Post-reform vs. Pre-reform	0.0043	0.00129	3.35	−10.2	1.88	−5.43
Mid-transition vs. Pre-reform	0.0046	0.00124	3.71	−12.1	1.63	−7.44
Late-transition vs. Pre-reform	0.0101	0.00139	7.24	−17.0	1.62	−10.50
Post-transition vs. Pre-reform	0.0118	0.00136	8.64	−19.3	1.53	−12.64

Notes: N = 26 for Pre-reform, 28 for post-reform, 8 for mid- late- and post-transition cohorts; for statistical significance at the 1% level the t-statistic is 2.896 for N = 8, and 2.485 for N = 25

Table 7.3 Pre-reform vs. post-reform $\hat{\beta}_i$ and $\hat{\alpha}_i$, circulatory system diseases

	$E[\hat{\beta}_i \mid r_i]$	$\hat{\sigma}_i^2$	$\hat{\sigma}_i^2/N_i$	$E[\hat{\alpha}_i \mid r_i]$	$\hat{\sigma}_i^2$	$\hat{\sigma}_i^2/N_i$
Pre-reform cohorts, $r_i = 0$	0.0787	0.000116	0.0000004	7.33	1.8	0.07
Post-reform cohorts, $r_i = 1$	0.0989	0.0000502	0.0000018	2.11	1.6	0.06
Mid-transition cohorts, $r_i = 1$	0.1045	0.0000021	0.0000003	1.31	0.02	0.003
Late-transition cohorts, $r_i = 1$	0.1040	0.0000083	0.0000010	1.11	0.04	0.01
Post-transition cohorts, $r_i = 1$	0.1094	0.0000347	0.0000043	0.56	0.05	0.01
	$\overline{\hat{\beta}_i} - \overline{\hat{\beta}_0}$	Φ	t-statistic	$\overline{\hat{\alpha}_i} - \overline{\hat{\alpha}_0}$	Φ	t-statistic
Post-reform vs. Pre-reform	0.0202	0.0015	13.5	−5.23	0.355	−14.7
Mid-transition vs. Pre-reform	0.0257	0.0008	30.5	−6.02	0.269	−22.4
Late-transition vs. Pre-reform	0.0252	0.0012	20.7	−6.22	0.274	−22.7
Post-transition vs. Pre-reform	0.0307	0.0022	14.0	−6.77	0.275	−24.6

Notes: N = 26 for Pre-reform, 28 for post-reform, 8 for mid- late- and post-transition cohorts; for statistical significance at the 1% level the t-statistic is 2.896 for N = 8, and 2.485 for N = 25

Table 7.4 Pre-reform vs. post-reform $\hat{\beta}_i$ and $\hat{\alpha}_i$, nervous system diseases

	$E[\hat{\beta}_i \mid r_i]$	$\hat{\sigma}_i^2$	$\hat{\sigma}_i^2/N_i$	$E[\hat{\alpha}_i \mid r_i]$	$\hat{\sigma}_i^2$	$\hat{\sigma}_i^2/N_i$
Pre-reform cohorts, $r_i = 0$	0.0500	0.0002181	0.0000084	8.10	12.3	0.47
Post-reform cohorts, $r_i = 1$	0.0346	0.0001451	0.0000052	4.27	6.30	0.22
Mid-transition cohorts, $r_i = 1$	0.0282	0.0000088	0.0000011	4.02	0.711	0.089
Late-transition cohorts, $r_i = 1$	0.0518	0.0000444	0.0000055	1.03	0.281	0.035
Post-transition cohorts, $r_i = 1$	0.0603	0.0000032	0.0000004	0.49	0.003	0.004
	$\overline{\hat{\beta}_i} - \overline{\hat{\beta}_0}$	Φ	t-statistic	$\overline{\hat{\alpha}_i} - \overline{\hat{\alpha}_0}$	Φ	t-statistic
Post-reform vs. Pre-reform	−0.0153	0.00368	−4.20	−3.83	0.83	−4.59
Mid-transition vs. Pre-reform	−0.0218	0.00308	−7.10	−4.08	0.75	−5.45
Late-transition vs. Pre-reform	0.0019	0.00373	0.50	−7.07	0.71	−9.92
Post-transition vs. Pre-reform	0.0103	0.02955	3.50	−7.61	0.69	−11.07

Notes: N = 26 for Pre-reform, 28 for post-reform, 8 for mid- late- and post-transition cohorts; for statistical significance at the 1% level the t-statistic is 2.896 for N = 8, and 2.485 for N = 25

Table 7.5 Pre-reform vs. post-reform $\hat{\beta}_i$ and $\hat{\alpha}_i$, digestive system diseases

	$E[\hat{\beta}_i \mid r_i]$	$\hat{\sigma}_i^2$	$\hat{\sigma}_i^2/N_i$	$E[\hat{\alpha}_i \mid r_i]$	$\hat{\sigma}_i^2$	$\hat{\sigma}_i^2/N_i$
Pre-reform cohorts, $r_i = 0$	0.5310	0.0006681	0.0000026	7.98	5.24	0.20
Post-reform cohorts, $r_i = 1$	0.0533	0.0002216	0.0000079	3.92	5.96	0.21
Mid-transition cohorts, $r_i = 1$	0.0434	0.0000267	0.0000033	4.16	2.01	0.25
Late-transition cohorts, $r_i = 1$	0.0739	0.0001382	0.0000173	0.59	0.19	0.024
Post-transition cohorts, $r_i = 1$	0.0893	0.0000068	0.0000008	0.20	0.0004	0.00006
	$\overline{\hat{\beta}_i} - \overline{\hat{\beta}_0}$	Φ	t-statistic	$\overline{\hat{\alpha}_i} - \overline{\hat{\alpha}_0}$	Φ	t-statistic
Post-reform vs. Pre-reform	0.0002	0.00325	0.07	−4.06	0.64	−6.31
Mid-transition vs. Pre-reform	−0.0096	0.00244	−3.95	−3.82	0.67	−5.68
Late-transition vs. Pre-reform	0.0208	0.00445	4.66	7.39	0.48	−15.55
Post-transition vs. Pre-reform	0.0362	0.00186	19.48	−7.79	0.45	−17.3

Notes: N = 26 for Pre-reform, 28 for post-reform, 8 for mid- late- and post-transition cohorts; for statistical significance at the 1% level the t-statistic is 2.896 for N = 8, and 2.485 for N = 25

Table 7.6 Pre-reform vs. post-reform $\hat{\beta}_i$ and $\hat{\alpha}_i$, musculoskeletal system diseases

	$E[\hat{\beta}_i \mid r_i]$	$\hat{\sigma}_i^2$	$\hat{\sigma}_i^2/N_i$	$E[\hat{\alpha}_i \mid r_i]$	$\hat{\sigma}_i^2$	$\hat{\sigma}_i^2/N_i$
Pre-reform cohorts, $r_i = 0$	0.0456	0.0001049	0.0000040	1.86	1.10	0.04
Post-reform cohorts, $r_i = 1$	0.0720	0.0003497	0.0000125	0.17	0.01	0.0004
Mid-transition cohorts, $r_i = 1$	0.0577	0.0000166	0.0000021	0.19	0.002	0.00025
Late-transition cohorts, $r_i = 1$	0.0951	0.0003404	0.0000126	0.04	0.0023	0.00029
Post-transition cohorts, $r_i = 1$	0.0880	0.0000912	0.0000114	0.03	0.001	0.00002
	$\overline{\hat{\beta}_i} - \overline{\hat{\beta}_0}$	Φ	t-statistic	$\overline{\hat{\alpha}_i} - \overline{\hat{\alpha}_0}$	Φ	t-statistic
Post-reform vs. Pre-reform	0.0264	0.0041	6.50	−1.69	0.203	−8.33
Mid-transition vs. Pre-reform	0.0120	0.0025	4.87	−1.67	0.203	−8.25
Late-transition vs. Pre-reform	0.0495	0.0068	7.25	−1.83	0.203	−9.00
Post-transition vs. Pre-reform	0.0423	0.0039	10.77	−1.84	0.202	−9.10

Notes: N = 26 for Pre-reform, 28 for post-reform, 8 for mid- late- and post-transition cohorts; for statistical significance at the 1% level the t-statistic is 2.896 for N = 8, and 2.485 for N = 25

Table 7.7 Pre-reform vs. post-reform $\hat{\beta}_i$ and $\hat{\alpha}_i$, neoplasms

	$E[\hat{\beta}_i \mid r_i]$	$\hat{\sigma}_i^2$	$\hat{\sigma}_i^2/N_i$	$E[\hat{\alpha}_i \mid r_i]$	$\hat{\sigma}_i^2$	$\hat{\sigma}_i^2/N_i$
Pre-reform cohorts, $r_i=0$	0.0945	0.000044	0.0000017	1.03	0.05	0.002
Post-reform cohorts, $r_i=1$	0.0955	0.000005	0.0000002	1.45	0.03	0.001
Mid-transition cohorts, $r_i=1$	0.0968	0.000001	0.0000001	1.36	0.006	0.0008
Late-transition cohorts, $r_i=1$	0.0969	0.000004	0.0000005	1.35	0.021	0.003
Post-transition cohorts, $r_i=1$	0.0988	0.000004	0.0000005	1.04	0.015	0.002
	$\overline{\hat{\beta}_i} - \overline{\hat{\beta}_0}$	Φ	t-statistic	$\overline{\hat{\alpha}_i} - \overline{\hat{\alpha}_0}$	Φ	t-statistic
Post-reform vs. Pre-reform	0.0010	0.0014	0.75	0.43	0.05	8.15
Mid-transition vs. Pre-reform	0.0024	0.0013	1.75	0.33	0.05	6.49
Late-transition vs. Pre-reform	0.0025	0.0015	1.67	0.33	0.07	4.92
Post-transition vs. Pre-reform	0.0044	0.0015	2.93	0.01	0.06	0.21

Notes: N = 26 for Pre-reform, 28 for post-reform, 8 for mid- late- and post-transition cohorts; for statistical significance at the 1% level the t-statistic is 2.896 for N = 8, and 2.485 for N = 25

potential, however, that one-time flattening of the slope stands for drastic improvement. In the top-left segment, that one-time jump is followed by the usual pattern where the slopes of subsequent sets became steeper as the profiles shifted down.

Compared to pre-reform children, the ones that grew up toward the end of the nineteenth century and early twentieth century did sizably better as they may have been less exposed to hazards that have long-term effects on the nervous system. A similar, but more limited, exception appears for the diseases of the digestive system in the mid-transition cohorts. Both sub-categories likely received a big initial boost from laws that targeted food and water children consumed.

Meanwhile, the case for neoplasms in Table 7.7 reaffirms the results in earlier chapters that cancers seem impervious to regime change since the nineteenth century, suggesting that their links with childhood development may be tenuous. The inference, however, is restricted to the sub-category in aggregate, for there could have been types of cancers that had improved. Furthermore, as the appendix points out, the confidence attached to neoplasms is the least among all sub-categories. Some of the differences across the cohorts could be attributable to measurement issues stemming from the series of the sub-categories being not perfectly consistent across the ICD lists. The only category for which there can be most confidence is the aggregate non-communicable diseases. The remaining chapters will focus only on the aggregate non-communicable diseases.

7.3 Though the Internal Validity of the Results Is Poor

The main drawback of the results from the natural experiment, something true for most such experiments, is that they disallow good internally-valid comparisons. Comparisons would be internally valid only if the units being observed are identical or homogenous in every way except the exposure to reform in childhood. That is a very unrealistic demand to meet, especially when the underlying change is fundamental, with ramifications spanning decades.

Further, the framework does not address the issue how variables that reflect children's development may have swayed both parameters. The only way the childhoods differ in this natural experiment is whether they were pre- or post-reform. Further, it does not address issues such as the relative contributions of childhood influences and of the life-course events. While the reforms may have triggered a fundamental change in the conditions children received while growing up, it seems very unlikely the cohort-profiles were what they were solely because their childhood years were independent of the reforms or the only dimension along which the cohorts had differed was whether their childhoods were exposed to reforms. For instance, 'pre-reform' and 'post-reform' may be good guideposts for the historical direction of infectious diseases in childhood. That does not, however, uncover how the changing rates of childhood infectious diseases, themselves, had influenced non-communicable diseases later on. Chapter 8 explores this in some detail.

Nonetheless, the results do suggest that political factors can change the milieu in which children grow up. Even though political variables may not be as amenable to regular measurement as economic or epidemiologic variables, they may influence the economic and epidemiologic issues that tell on children's development.

One potential strength of the results may be their external validity. They indicate that fundamental political change, provided it also triggers policy changes in the arena of public health (relating especially to children's development), may result in similar types of broader change in aging. Moreover, they suggest that healthcare spending at the aggregate-level may have strong political roots. The political reforms opened up a way for people to reveal their preferences by voting, enabling policymakers to vie for the types of policies the masses may have preferred. Chapter 9 reconnects with this issue while discussing the role of income elasticity of healthcare spending, deducing that the elasticity may not be construed in the usual sense if the allocations to healthcare have strong links to the political sphere.

References

Angirst, J. D., & Pischke, J.-S. (2009). *Mostly harmless econometrics, an empiricist's companion*. Princeton: Princeton University Press.

Heckman, J. J. (2008). *Econometric causality* (IZA Discussion Papers, No. 3425). http://nbn-resolving.de/urn:nbn:de:101:1-2008040906

Wooldridge, J. M. (2009). *Econometric analysis of cross section and panel data*. Boston: MIT Press.

Chapter 8
The Signals from the Childhood Years

8.1 When Initial States, Too, May Tilt the Slope

Factors that sway the initial state α_i of cohort i's profile likely do so well before they can tell on its slope β_i that forms until decades later. Further, the finiteness of lifespan implies that α_i and β_i could relate inversely, and α_i, on its own, may be one of the factors that tilt β_i mechanically. For those reasons it is best to begin the analysis with α_i before figuring its proper role in the formation of β_i. Consider the following linear equation for α_i:

$$\alpha_i = \mu_\alpha + \rho c_i + \eta_i. \tag{8.1}$$

Here, μ_α is a constant, c_i for cohort i is its childhood regime's *realized* value that influences the initial state, and η_i is a mean-zero error. However, any value of c_i indicated by a proxy for the childhood years is very likely to correlate with the error, and that needs to be amended as much as possible for estimating the relation of c_i and α_i accurately. Suppose the error η_i were a linear function of various other traits X_i of the regimes, and of an error term v_i

$$\eta_i = X_i^{'}\gamma + v_i. \tag{8.2}$$

As v_i and X_i are uncorrelated by design, the expectation of Eq. 8.2 is $E[\eta_i|X_i] = X_i^{'}\gamma$. Taking expectation on both sides of Eq. 8.1 and plugging this result into the equation yields

$$E[\alpha_i|X_i, c_i] = E[\alpha_i|X_i] = \mu_\alpha + \rho c_i + E[\eta_i|X_i]$$

S. Arora, *The Transitions of Aging*, International Perspectives on Aging 12,
DOI 10.1007/978-3-319-14403-0_8

or

$$\alpha_i = \mu_\alpha + \rho c_i + X_i^{'}\gamma + \omega_i, \tag{8.3}$$

where the coefficient of interest is ρ; and ω_i is a residual, a leftover after any influence of c_i and X_i on α_i has been purged. The accuracy of ρ still rests on the notion that the *observed* controls X_i are the only reason η_i and c_i correlate. Omitted variables undercut the notion, which is true for all realistic scenarios (Angirst and Pischke 2009).

Meanwhile, the baseline setup for β_i would be one that first takes into account any mechanical influence of the initial state α_i, as in the following linear set up

$$\beta_i = \mu_\beta + \acute{\theta}_\alpha \alpha_i + \epsilon_i \tag{8.4}$$

where the estimates of $\acute{\theta}_\alpha$ should be negative given that lifespan is finite. Now, the childhood regime has to be brought into the picture. However, some of it is already there in the guise of α_i as stated in Eq. 8.3. If α_i and β_i have at least one influence from childhood in common, then inserting α_i in Eq. 8.4 along with the childhood variable would amount to letting in the influence of that variable twice, one time in the guise of α_i, and the second time on its own. A reasonable way to tackle the issue is to use α_i 'purged' of any influence of the childhood variables. That can be done by using the estimated residuals $\widehat{\omega}_i$ from Eq. 8.3, so that

$$\beta_i = \mu_\beta + \theta_\alpha \widehat{\omega}_i + \epsilon_i \tag{8.5}$$

and then by bringing in the childhood regime variable

$$\beta_i = \mu_\beta + \theta_\alpha \widehat{\omega}_i + \delta c_i + \varepsilon_i, \tag{8.6}$$

and finally evaluating the results in the presence of various other macro variables or controls X_i

$$\beta_i = \mu_\beta + \theta_\alpha \widehat{\omega}_i + \delta c_i + X_i^{'}\gamma + \pi_i \tag{8.7}$$

where π_i is an error term and δ is the coefficient of interest. The empirical results will be presented in the following sequence:

1. For α_i, estimate: (i) the baseline with only the childhood regime's proxy, Eq. 8.1; (ii) the extended version, Eq. 8.3, with other variables.
2. For β_i: (i) the baseline with $\widehat{\omega}$, Eq. 8.5; (ii) then bring in the childhood proxy, as in Eq. 8.6; and finally, Eq. 8.7 that has $\widehat{\omega}_i$, the proxy, and other variables.

The estimation method is Ordinary Least Squares with standard errors corrected for heteroskedasticity and serial correlation by using the Newey-West/Bartlett window with two lags.

8.2 As Could Proxies for the Childhood Years

Listed in Table 8.1, the first proxy is average stature. Chapter 6 pointed out that average stature is likely a more incisive instrument than broader gauges of inequality. If inequality in the childhood years matters to aging over adulthood, a measure should reflect: (1) how the inequality affected only childhood development; and (2) how the inequality from then, and then only, was channeled to the profiles defined over the adult years. Average stature is shaped only by what goes on in the childhood years; and the influences from then get isolated since stature does not grow after the onset of adulthood. It is then channeled into the profiles over the adult years.

Stature's expected relation to β_i is positive: all else constant, as children on average fare better, average stature of a cohort grows, and the rates of non-communicable diseases in the profiles may diminish as the cohort ages, resulting in larger β_i. Meanwhile, its expected relation with α_i is negative: taller average stature, sturdier initial conditions, less elevated profile at the very start, smaller α_i.

The second childhood proxy gauges infectious disease as a proportion of all disease-specific causes among 0–14-year olds when the cohorts were 0–14 years of age. All else equal, poorer epidemiologic settings would have a greater proportion of infectious disease, more chances of setbacks to children's development, making it likelier for higher rates of non-communicable disease to appear in the adult years, fostering flatter profiles and smaller β_i. The expected relation with α_i is positive: poorer epidemiologic setting, larger fraction of infectious diseases, greater risks to children's growth, more elevated the profile from the get go, larger α_i.

The third variable, measured in each birth decade, is the average age-standardized rates of complications of pregnancy and childbirth, seen as an indicator of stresses women may have faced in the decade. Besides the link between in-utero stresses and adult-age non-communicable diseases outlined in Barker (1992, 1994, 1998), this proxy may highlight how gender inequity may be transmitted to high rates of non-communicable diseases years later (Osmani and Sen 2003). It is vital to note that mechanisms outlined in the fetal-origins hypothesis have been questioned and are not universally accepted (e.g. Lancet 2001), and scholars have emphasized the role of life-course influences (Ben-Shlomo and Davey Smith 1991; Davey Smith et al. 2001; Ben-Shlomo and Kuh 2002; Kuh et al. 2002; Kuh and Davey Smith 2004).

The results, as will become clearer later in this chapter, indicate that for all childhood proxies there is a childhood-linked and a life-course-linked segment in the rates of non-communicable diseases. The more critical issue for this proxy is whether it reflects in-utero stresses or reflects the nutritional status of women during pregnancy, which, in turn, could stand for the nutritional status of women of child-bearing age in general (e.g. Forsen et al. 1997). The aggregate-level analysis is incapable of distinguishing one from the other as those traits are likely to correlate positively.

Table 8.1 Childhood proxies and macro variables

	Variable	Measurement	Coefficient
	Childhood proxies (observations)		
1	Average adult stature (two sets, N = 31 each)	Centimeters, at age 18	δ
2	Infectious diseases at age 0–14 (N = 50)	(Infectious diseases $\div$ all diseases) in 0–14-year olds when cohorts were 0–14 years of age	δ
3	Complications of pregnancy and childbirth (N = 44)	Average age-standardized rate in each birth decade	δ
	Macro variables		
1	The 'purged' α_i (only for studying β_i)	Estimated residuals $\widehat{\omega}_i$ from Eq. 8.3	θ_α
2	Functional distribution of national income	(Labor income $\div$ capital income) in birth decade	γ_1
3	Risk-adjusted real per capita income in the childhood years	(Average real per capita income $\div$ standard deviation of per capita income) in the childhood years	γ_2

All else constant, in each cohort's birth-decade, better the status of women during pregnancy, lesser the likelihood of complications, better the outcomes for children from the very start, smaller the rates of non-communicable diseases in the profile years later, making the profile steeper and β_i larger. Its expected relation to α_i would be positive: cohorts born when the stresses are less are likelier to start better at the onset of their profiles.

Further, it helps to keep in mind that any proxy is a stand-in for the childhood regime, not the regime itself. There could be other factors in the regimes that are yet unknown, or cannot be measured. For instance, germs were not clearly linked to various diseases until the last third of the nineteenth century or so. That knowledge, or the lack of it, would be a telling feature of the regimes, but it is also one that is very difficult to isolate: when and to what extent the public knew about the linkages and had changed behavior, irrespective and independently of public policy, may not be amenable to measurement at the aggregate level.

Moreover, for two main reasons, each proxy will be examined separately. It will be clear by the end of the chapter that they likely have an unsymmetrical relation to various diseases comprising the aggregate, indicating that no childhood variable should be written off or be seen as more important than others. The evidence indicates, instead, that the childhood regime is likely inadequately captured by even these three proxies, and more work may be required to narrow the gaps.

Furthermore, the data on the proxies span different periods. The data on stature begin in the early nineteenth century, but there is only one series. It will be used twice, once with the first set of cohorts (I and II) and then for the second set (III and IV); each set has only 31 observations. Meanwhile, data on infections and on pregnancy complications begin toward the mid-nineteenth century, yielding about 45–50 observations; and the disadvantage is that cohorts from the first three to four decades of the nineteenth century have to be left out, a loss of 12–16 Malthusian cohorts from the analysis.

8.3 Alongside Macro Variables

The macro variables to be used alongside the childhood proxies are: (1) risk-adjusted real per capita income in the childhood years; (2) the functional distribution of national income in the birth decades. Besides acting as controls for the proxies, they bring a narrative of their own and may have played distinct roles in aging.

Before seeing how they are measured, first consider the proper role of a control in the outlined set up. The purpose of a control is to check if something else in the childhood years, something other than the proxies, is not the source of the results. For that reason it is critical that the control pertain to the childhood years as well.

Further, the control should not have been the outcome itself (Angirst and Pischke 2009). One of the outcomes is β_i, which is determined by the profile's shape over age 25–74. Any variable built up from the data already in the profiles is not useful and will be an ineffectual control. That includes indicators such as period-life expectancies, probabilities of death, cohort-life expectancies, among others. They all, too, could have been the outcome. The calculation of cohort-life expectancies at age 25, for instance, uses data on deaths from age 25 onward in the cohort's entire life path. Some of that data is already in the profile. Using the expectancy as a control would amount to using an outcome to control for the outcome. Lastly, the control should be amenable to measurement at the aggregate level.

The bottom half of Table 8.1 lists the macro variables that also act as controls for the childhood proxies. The first one to be used only for β_i is the 'purged' α_i as discussed in Sect. 8.1; it is expected to relate inversely to β_i as lifespans for all cohorts must be finite.

The second one reflects the trends in income and wealth distribution since the nineteenth century: the shares of national income going to labor and to capital, a mainstay of aggregate-level models. Recommending its use is the historical fact that a growing long-term share of capital incomes has coincided with widening inequality, and with steeper socio-economic gradients, because the ownership of capital has usually been concentrated in the upper rungs of the socio-economic ladder: greater the concentration of capital ownership, higher the share of national income going to capital, larger the share of incomes going to the upper rungs, smaller the LK ratio (Piketty 2014; Lindert 2000). In that sense, a falling LK ratio reflects widening inequality. Besides possibly influencing the profiles on its own, the ratio helps hone stature's role by checking that it is not usurping the influence of inequality in general, which can manifest in many ways other than children's physiological growths.

Moreover, over the very long term, trends of income inequality may have coincided with changes in the polity and in the new social policies that were adopted. For instance, policies that initiated State-supported schooling, and later made primary and secondary public schooling universal, are likely to have boosted people's awareness, knowledge and skill (Harris 2004; Lindert 2004). In this study, however, schooling is an omitted variable, mainly because the data on it are insufficient and poor. Controlling for the trends in the LK ratio may, however, serve as a stand-in for the long-term omitted issues such as schooling that the changing polities may have fostered.

The risk-adjusted real per capita income is measured as the average real per capita income over the childhood years adjusted for its standard deviation then. Growing real incomes signify better access to material resources, likely giving successive cohorts of children the means to a better start. However, real incomes may have been more unsteady in the childhood years of some cohorts than of others, such as for the cohorts brought up in the inter-war years. That could have meant more setbacks for at least some segments of children within those cohorts, contributing years later to smaller β_i; hence the adjustment for the volatility.

8.4 Average Stature Had a Long Reach

Table 8.2 has the results of the relation of stature to α_i. The top half has results for cohorts in sets I and II; the bottom half, for sets III and IV. The column 'Eq. 8.1' shows that stature related inversely to α_i, posting an adjusted R-square of 60–64%.

Table 8.2 Results for α_i, average stature as childhood proxy

	Eq. 8.1	Eq. 8.3	Eq. 8.3
Set I & II			
$\hat{\mu}_\alpha$	364.7	244.7	243.6
	(44.3)	(51.5)	(48.8)
$\hat{\rho}$	−2.10	−1.23	−1.20
	(0.26)	(0.33)	(0.31)
$\hat{\gamma}_1$		−12.52	−15.05
		(3.10)	(3.11)
$\hat{\gamma}_2$			−0.27
			(0.09)
$\overline{R}^2$	64%	75%	79%
Set III & IV			
$\hat{\mu}_\alpha$	392.6	230.5	232.9
	(55.8)	(60.9)	(60.6)
$\hat{\rho}$	−2.24	−1.10	−1.10
	(0.33)	(0.38)	(0.37)
$\hat{\gamma}_1$		−16.7	−18.7
		(3.10)	(3.4)
$\hat{\gamma}_2$			−0.25
			(0.11)
$\overline{R}^2$	61%	77%	79%

Note: N= 31 for Set I & II, and for Set II & IV. Standard errors are in parentheses

The next column brings in the functional distribution to find that it too had influenced α_i, reducing the coefficient on stature, and raising the R-square by about 11–16 percentage points. Between stature and LK, the results indicate that more inequality in the childhood years worsens the initial states, elevating the profiles since early in the adult years. The coefficient on the risk-adjusted real per capita income in the last column indicates that economic prosperity in the childhood years may have helped shift the profiles down as well, though the size of the coefficient is relatively small. Between them they explain about 80% of the variation in α_i across cohorts.

Meanwhile, in Table 8.3, column 'Eq. 8.5' shows that α_i, 'purged' of any influence of the three variables used in Table 8.2, relates inversely to the slopes. While the R-square is merely 7–10%, looking across the columns it is clear that the relation turns out quite resilient. It indicates there is most likely a mechanical link: a lower initial state alone makes the slope somewhat steeper since there is a limit to the lifespan.

Column 'Eq. 8.6' shows that stature was a strong contributing factor. Three notable things occur when it enters the picture: (1) the coefficient on the 'purged' α_i in the previous column endures nearly unscathed; (2) the R-square leaps from about 7–10% to 83–84%; and (3) the constant adjusts sizably. They suggest that the trends in stature are a good proxy for the changing childhood regimes: if children on

Table 8.3 Results for β_i, average stature as childhood proxy

	Eq. 8.5	Eq. 8.6	Eq. 8.7
Set I & II			
$\widehat{\mu}_\beta$	0.0780	−0.1608	−0.1034
	(0.0014)	(0.0202)	(0.0192)
$\widehat{\theta}_\alpha$	−0.0005	−0.0005	−0.0005
	(0.0002)	(0.0001)	(0.00004)
$\widehat{\delta}$		0.0014	0.0010
		(0.0001)	(0.0001)
$\widehat{\gamma}_1$			0.0069
			(0.0014)
$\widehat{\gamma}_2$			0.0001
			(0.00004)
$\overline{R}^2$	10%	83%	91%
Set III & IV			
$\widehat{\mu}_\beta$	0.0770	−0.1752	−0.1065
	(0.0015)	(0.0205)	(0.0183)
$\widehat{\theta}_\alpha$	−0.0004	−0.0004	−0.0004
	(0.0002)	(0.0001)	(0.0001)
$\widehat{\delta}$		0.0015	0.0010
		(0.0001)	(0.0001)
$\widehat{\gamma}_1$			0.0077
			(0.0010)
$\widehat{\gamma}_2$			0.00007
			(0.00003)
$\overline{R}^2$	71%	84%	92%

Notes: N = 31 for Set I & II, and for Set II & IV. Standard errors are in parentheses

average were ending up stunted, then it must be that the regime was deteriorating; and vice versa. A constant would describe such scenarios poorly, and would have to adjust in the presence of a variable that captures them better.

Between Tables 8.2 and 8.3, stature likely swayed the profiles in two ways: (1) it helped shift the initial state down, more than the 'purged' α_i, steering the profiles to a better start; (2) it travelled up the profiles, channeling the social gradient from childhood to β_i. This double role can be defined as the *cohort effect*: it sets the cohorts on different paths at the very start, and then lingers to shape the paths until years later.

The column 'Eq. 8.7' looks into the matter that stature could be merely picking up the effects of broader trends in income and wealth inequalities. In the presence of the functional distribution, the coefficient on stature reduces. Moreover, the R-square increases, though by much less than the leap it made in the previous column.

At this point, it helps to briefly reconnect with the discussion in Chap. 6. The results pertain to aggregated profiles, but they have at least three implications for how the profiles of broad social classes within the aggregate could have changed. The premise was that since the elite had reached close to modern heights early in the nineteenth century, the heights of the non-elite in each cohort likely swayed the trends of the overall average. As the average grew, the aggregated profiles shift down and became steeper. If true, three within-profile or within-cohort changes could have occurred: (1) The non-elites' α_i may have fallen more, and have come closer to the elites' α_i; (2) compared to the past, gaps in their β_i may have narrowed as the β_i of the non-elite increases toward that of the elite. Holding their average statures constant, unsymmetrical access to medical care, especially when health insurance was not universal, may have had the reverse effect: with better access, the β_i of the elite profiles would grow before the ones of the non-elite profiles; the gap may shrink as the medical advances become more widely available.

A third within-cohort change could be that the observed average height of a specific cohort could have 'grown' over the term of its profile. As the non-elite on average were shorter, all else constant, a larger fraction of them may presumably perish sooner, or age less. Consequently, over the 50-year term of each profile, the average of the residual height distribution may increase. This tendency may be less pronounced more recently than in the nineteenth century when heights across the classes varied more, and the access to insurance or medical care was likely much less uniform. Studying these intra-profile or within-profile movements is beyond the scope of this book.

In the next chapter, to study the role of aging in healthcare, one needs to extract the childhood-linked signals implied by Tables 8.2 and 8.3. The first step toward the goal is to find out how well the signals help track the original cohort parameters, the ones that had simulated the actual rates very closely in Chap. 4. This is done in two stages: (1) combine the estimated coefficients in Tables 8.2 and 8.3 with each cohort's fundamentals (stature, LK ratio etc.), and generate the implied parameters for the profiles; (2) compare the implied parameters to the original ones.

Figure 8.1a shows the implied and the original α_i of cohorts, each one ordered by the middle year of the birth decades on the horizontal axis. They are calculated by using the coefficients in last column of Table 8.2. The correlation between the implied and the originals is 0.88. The mean of the squared percentage errors is 3.8%. Reassuringly, the implied α_i track the long-term trends well: after falling during the first two decades of the nineteenth century, they grew for cohorts in the mid-nineteenth century, eventually falling again for the transition and the post-transition cohorts.

However, their trend is more subdued than that of the originals. They are unable to capture the severe situation of the mid-nineteenth century cohorts; and somewhat understate the progress in the early twentieth century cohorts (higher α is worse). In contrast, over the post-transition stage they are somewhat better than the originals. This last feature suggests that children in the post-transition era may have developed a new set of childhood-related issues these childhood variables are unable to capture.

Figure 8.1b shows the results for β_i. They are based on coefficients from the last column in Table 8.3. The implied β_i fare better than the implied α_i: their correlation to the actuals is 0.94; the mean of squared percentage errors 0.11%. They do better at picking up the severity of the situation in the mid-nineteenth century as

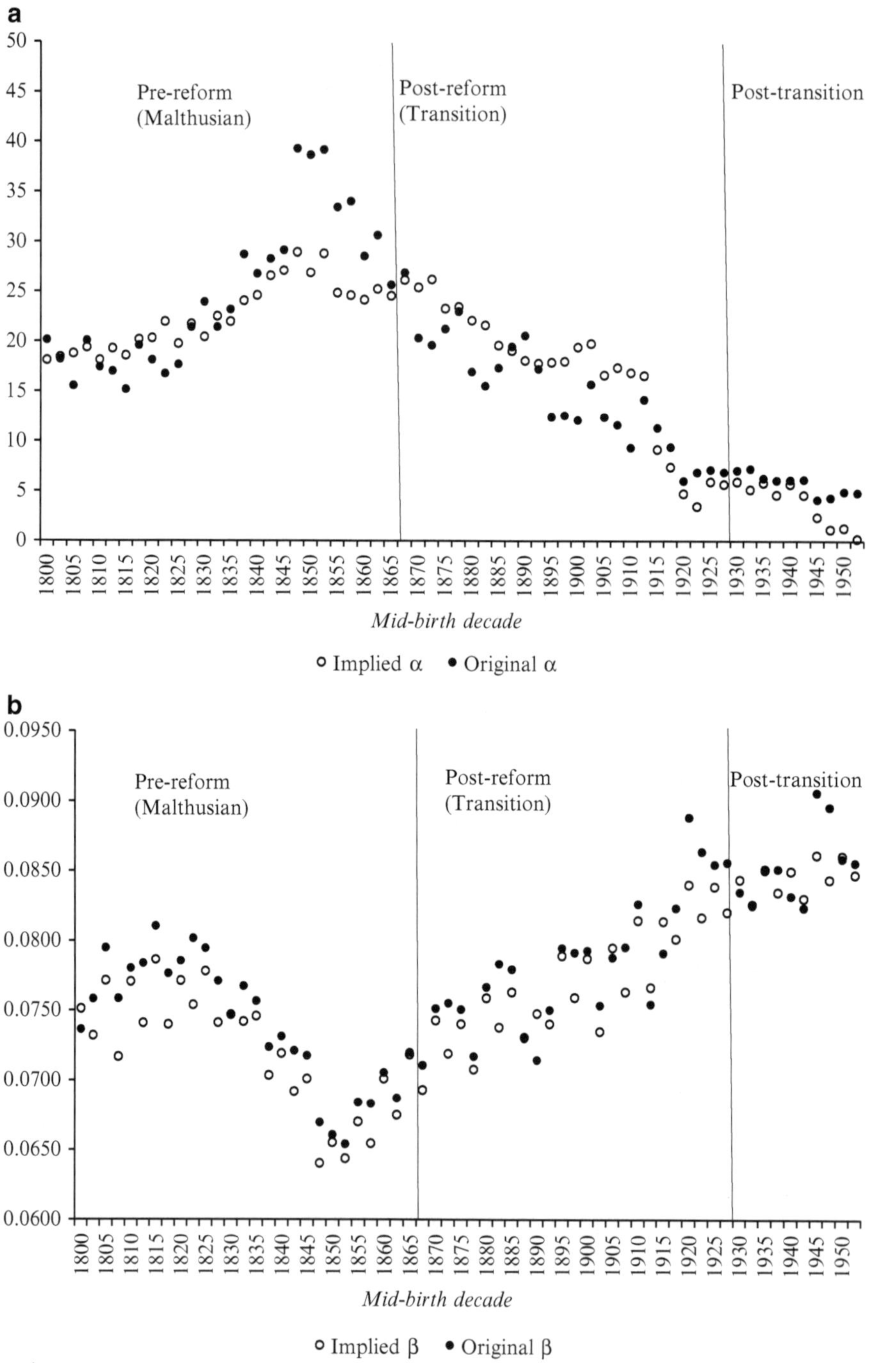

Fig. 8.1 Original cohort parameters vs. ones implied by average stature as childhood proxy (**a**) α_i (**b**) β_i (Source: author's calculations)

well as the main direction of the trends over the transition. They are mildly off in the post-transition stage, and in the early- to mid-nineteenth century, being less than the originals in both cases.

I will present such an exercise for each childhood proxy. A separate section later assembles the results, and examines how well the implied parameters help simulate the long-term path of the aggregate rates of non-communicable diseases, an exercise similar to the one conducted in Chap. 4.

8.5 As Did Infectious Disease in Childhood

The results in Table 8.4 indicate that larger the proportion of infectious diseases in children, more elevated the age-profiles from the outset. However, the coefficient shrinks perceptibly in the presence of other variables in the last column, and the R-square grows to about 80%. The coefficient $\widehat{\gamma}_1$ conveys that less income inequality in the birth decades associate with less elevated profiles, while $\widehat{\gamma}_2$ suggests that real incomes helped lower the initial states of the profiles as well.

Meanwhile, the second column of Table 8.5 indicates that the 'purged' initial state relates inversely to the slope, a relation that stays robust across the columns, but explains at most 13% of the total cross-cohort variation. The explanatory power in the next column rises sizably as the R-square leaps to about 63% with infectious diseases in the picture. The last column conveys that broader trends of income distribution during the birth decades have been influential, with the slopes becoming steeper as labor's share rises relative to that of capital. Together the variables explain about 91% of the variation.

Based on the coefficients in the last column of Tables 8.4 and 8.5, the first thing to note about the implied parameters in Fig. 8.2a, b (see page 143) is the shorter history than the one for stature. Nonetheless, the implied α_i correctly track the broader trends, but are unable to mimic the severity in the mid-nineteenth century. After being more or less in line with the actuals in the transition stage, they are more off the mark for cohorts in the inter-war years, particularly for ones from the 1930s;

Table 8.4 Results for α_i, infectious diseases as childhood proxy

	Eq. 8.1	Eq. 8.3	Eq. 8.3
$\widehat{\mu}_\alpha$	−14.9	61.9	66.9
	(3.3)	(6.8)	(6.5)
$\widehat{\rho}$	0.60	0.02	0.12
	(0.07)	(0.05)	(0.05)
$\widehat{\gamma}_1$		−24.7	−26.1
		(2.3)	(2.2)
$\widehat{\gamma}_2$			−0.56
			(0.12)
$\overline{R}^2$	36%	74%	79%

Notes: N = 50. Standard errors are in parentheses

Table 8.5 Results for β_i, infectious diseases as childhood proxy

	Eq. 8.5	Eq. 8.6	Eq. 8.7
$\widehat{\mu}_\beta$	0.0775	0.1000	0.0559
	(0.0014)	(0.0030)	(0.0051)
$\widehat{\theta}_\alpha$	−0.0005	−0.0005	−0.0005
	(0.0001)	(0.0001)	(0.00004)
$\widehat{\delta}$		−0.0004	−0.0001
		(0.00005)	(0.00005)
$\widehat{\gamma}_1$			0.0142
			(0.0014)
$\widehat{\gamma}_2$			0.0002
			(0.00003)
$\overline{R}^2$	13%	63%	92%

Notes: N = 50. Standard errors are in parentheses

they then improve more than the originals for the 1940s cohorts. Their correlation with the original parameters is 0.9, and the average percentage error is 4%, and the squared percentage error of 9.2% indicates some large mismatches.

The implied β_i, meanwhile, fare much better. They capture the acuteness of the situation for the mid-nineteenth century cohorts well; remain mostly within a narrow band of the actuals in the transition stage; but fare somewhat poorly in the post-transition stage where they are a lot less variable than the actuals. The correlation is 0.96; the mean of the squared percentage errors is 1.2%.

8.6 Pregnancy Complications Reached Long, Too

The results presented in Table 8.6 (on page 144) roughly follow the pattern of the other two proxies. More pregnancy complications in the birth decade associate with more elevated profiles, and that explains about 72% of the cross-cohort variation. However, the coefficient shrinks in the presence of controls while the adjusted R-square improves to about 78%.

Meanwhile, like the results for the other proxies, the coefficient on the 'purged' α_i in the first column of Table 8.7 (page 144) proves durable, but explains only a sliver of the cross-cohort variation of β_i. The R-squares leaps to 86% when the pregnancy complications are added to the picture, implying that, all else constant, more severe the rates of complications at the time the cohorts were born, flatter the profiles, and smaller the β_i. Introducing controls shrinks the coefficient, but the explanatory power improves to 92%.

Like the other two proxies, The implied α_i in Fig. 8.3a miss the severity of the situation for the mid-nineteenth century cohorts; track the broad trends well; though understate the improvement in the mid-to-late transition stage, more so for the 1930s cohorts than for others; while misgauge the 1940s cohorts sizably. The overall correlation with the actuals is 0.9, the average percentage error 1.3%.

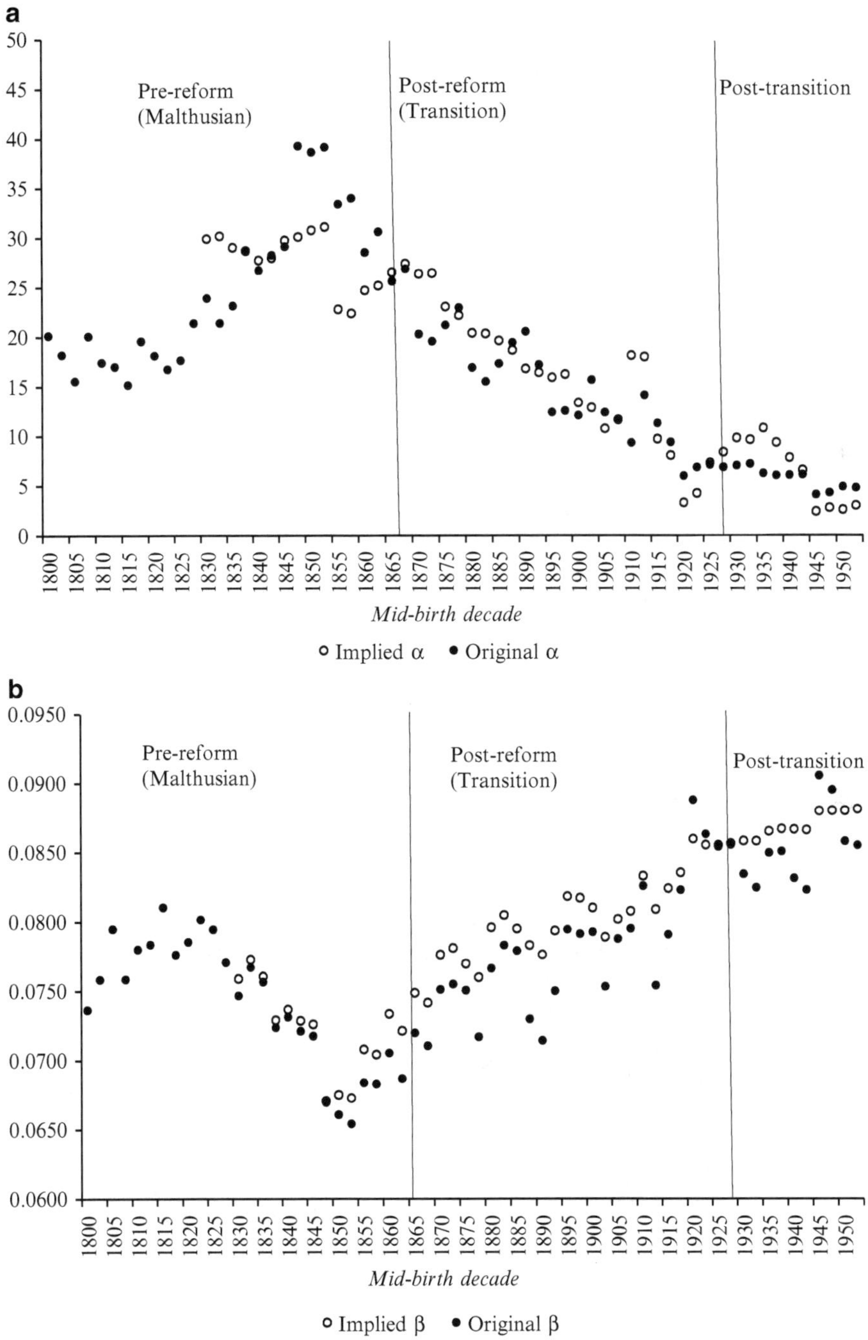

Fig. 8.2 Original cohort parameters vs. ones implied by infectious disease as childhood proxy (**a**) α_i (**b**) β_i (Source: author's calculations)

Table 8.6 Results for α_i, pregnancy complications as childhood proxy

	Eq. 8.1	Eq. 8.3	Eq. 8.3
$\widehat{\mu}_\alpha$	−2.22	39.4	46.9
	(1.1)	(13.8)	(12.8)
$\widehat{\rho}$	0.79	0.35	0.31
	(0.07)	(0.15)	(0.14)
$\widehat{\gamma}_1$		−16.5	−18.2
		(5.30)	(4.9)
$\widehat{\gamma}_2$			−0.12
			(0.04)
$\overline{R}^2$	72%	77%	78%

Notes: N = 44. Standard errors are in parentheses

Table 8.7 Results for β_i, pregnancy complications as childhood proxy

	Eq. 8.5	Eq. 8.6	Eq. 8.7
$\widehat{\mu}_\beta$	0.0779	0.0903	0.0596
	(0.0012)	(0.0010)	(0.0060)
$\widehat{\theta}_\alpha$	−0.0005	−0.0005	−0.0005
	(0.0001)	(0.00005)	(0.00004)
$\widehat{\delta}$		−0.0005	−0.00022
		(0.00004)	(0.00007)
$\widehat{\gamma}_1$			0.0118
			(0.0023)
$\widehat{\gamma}_2$			0.00003
			(0.00002)
$\overline{R}^2$	8%	86%	92%

Notes: N = 44. Standard errors are in parentheses

Meanwhile, the implied β_i in Fig. 8.3b capture the turning point in the mid-nineteenth century well; travel up closely with the actuals in the transition stage; but begin straying in the inter-war years and into the post-transition period where they are steadier than the actuals. In contrast to α_i, their more consistent performance ensures very low mean percentage error (−0.1%) and the squared percentage error is 0.1%.

8.7 They Yield Similar, but Distinct, Long-Reaching Signals

The common features in the results across the proxies are summarized in Table 8.8. Both implied parameters correlate strongly with the original ones. For the implied α_i, however, the percentage error is much larger than the ones for the implied β_i. It indicates that the proxies are unable to fully capture various aspects of the

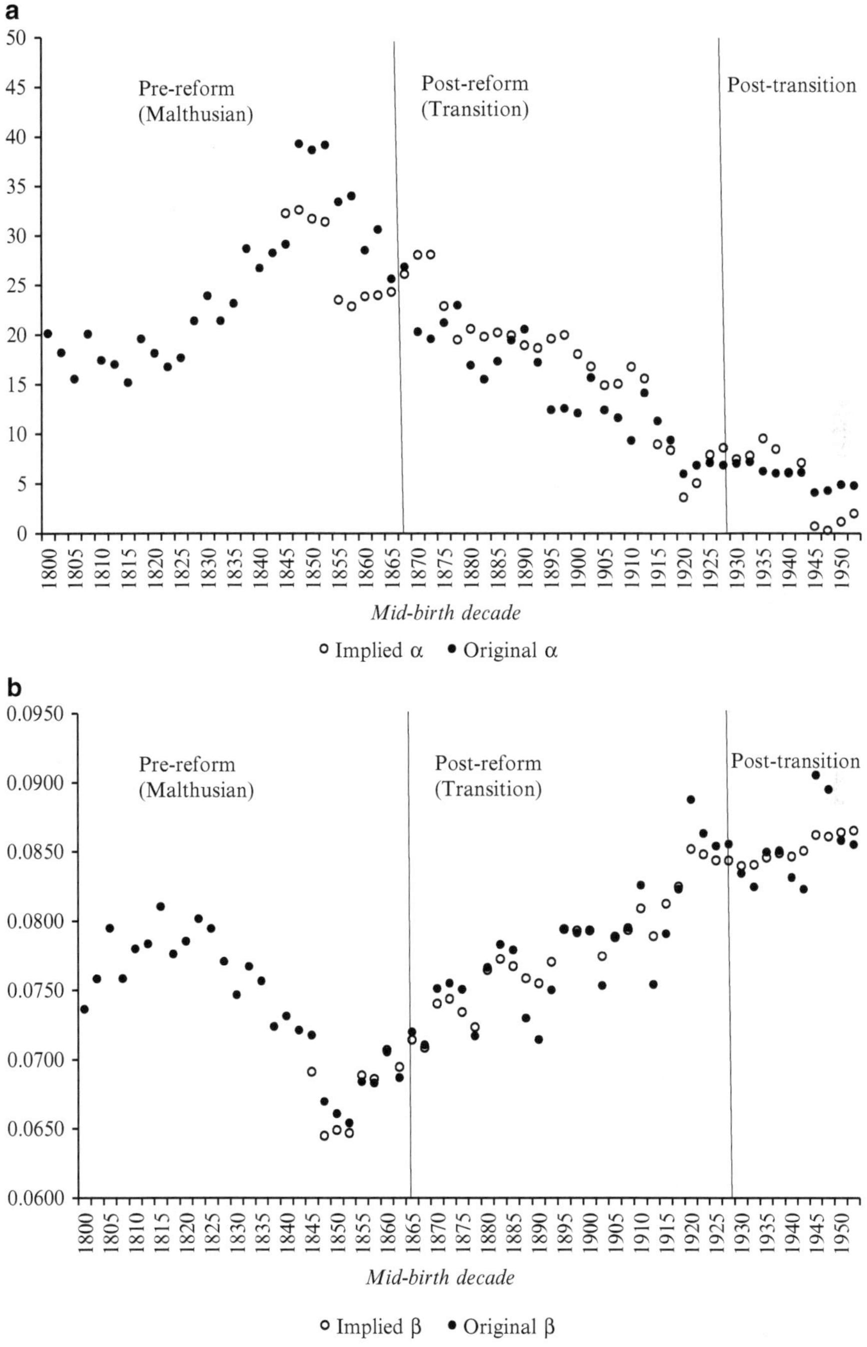

Fig. 8.3 Original cohort parameters vs. ones implied by pregnancy complications as childhood proxy (**a**) α_i (**b**) β_i (Source: author's calculations)

Table 8.8 Relation of the implied α_i, β_i to the original α_i, β_i

	Correlation	Mean E^{2*}	$\sqrt{MSE}$
Implied α_i vs. originals			
Average stature	0.88	3.8%	19.4%
Infectious diseases	0.90	9.2%	30.3%
Pregnancy complications	0.90	13.4%	36.6%
Implied β_i vs. originals			
Average stature	0.95	0.09%	3.0%
Infectious diseases	0.96	1.2%	3.5%
Pregnancy complications	0.87	0.1%	2.3%
Correlation of implied α_i, β_i			
Average stature	−0.91		
Infectious diseases	−0.90		
Pregnancy complications	−0.94		

Notes: N = 62 (stature); 50 (infectious disease); 44 (pregnancy complications)
*E is % error between the implied and the actuals; $\sqrt{MSE}$ is root mean square of E

childhood regime that tell on α_i, but they seem to do reasonably well at capturing issues that sway the original β_i.

Further, for all proxies, the implied α_i and β_i correlate inversely; Figs. 8.1, 8.2, and 8.3 strongly indicate that over the long term the two parameters trend in opposite directions. In addition, the trends of β_i relate less to the 'purged' α_i and more to socio-economic variables in the childhood years. The mechanical relation between the initial state and the slope, the one likely owing to the finiteness of lifespan, explains at most 7–13% of why β_i grew as α_i fell. The rest is likely because of socio-economic change, which is why studying historical contexts is vital for grasping modern aging.

Moreover, the proxies and the functional income distribution appear to be the main sources of what may be called the cohort effect: they sway the initial state as well as the slope. Meanwhile, role of the volatility-adjusted real per capita income in the childhood years appears to have been stronger for α_i than for β_i, indicating that it may not be the source of a cohort-effect as defined.

Nonetheless, all results are subject to the omitted variable bias, particularly for the implied α_i that seem to fare not as well as the implied β_i. Some amount of shortfall should be expected because the profiles are subject to many types of life-course events and developments, which, by definition, are beyond the reach of the childhood proxies. Another source of the shortfall is that the original parameters, too, had helped approximate the time series, even as the approximation was close. In the following discussion, I subsume any approximation errors as elements of the life-course segment.

For the matter at hand, however, the omitted variable issue comes with a twist: each proxy might be an omitted variable for the other. The bias would be positive, if the proxy being used relates positively to the omitted one; it is negative if they relate negatively. For instance, if infectious diseases were to relate negatively to stature, in the absence of the infectious disease variable, the estimate from the stature is likely to be too low; likewise for the estimate for the infectious disease proxy in the absence of stature. To use another example, if a regime of high infection rates is also one in which pregnancies are more perilous (a positive relation between the two), without infectious diseases, the estimates from the pregnancy proxy might be too high; likewise for the infections proxy. Logically, then, for the infections proxy, the estimates are too low without stature, and too high without pregnancy complications, and the net result uncertain.

The question then is: which variable should be relied upon? Each tells an important story about childhood: stunting, infections, the state of mothers during pregnancy and childbirth. How might the stories be combined quantitatively? There may not be a clear-cut answer, though there is one crude way: for each cohort, average the parameters implied by the three proxies.

Figure 8.4a, b show how the averaged α_i and β_i fare against the originals. The series in the early nineteenth century is dominated by stature for it is the only proxy available over the first three decades. All told, the general long-term pattern is similar to that of Figs. 8.1, 8.2, and 8.3.

To test the worth of these implied averaged parameters, the next question to ask is: how well do they simulate the long-term path of the aggregate rate? Between the two parameters, β_i is the more important one for studying the role of aging in healthcare spending in the following chapters. With that in mind, the question becomes: how well can the implied average β_i help simulate the long-term path of non-communicable diseases?

That can be explored by doing an exercise: arm each cohort with its own original α_i so that it is no longer the main source of any gap between the actual and the simulated results. Pair each cohort's original α_i with the cohort's average β_i implied by the proxies. These β_i will now be the main source of why the simulated rates would be off the actual ones. This exercise is identical to the one in Tables 4.1 and 4.2 that had used the original parameters to simulate the aggregate rates. The difference here is that each cohort's average β_i implied by the proxies is being used now instead of the original β_i.

Figure 8.5 (on page 149) shows that the averaged implied β_i do reasonably well at capturing the long-term trends. Their correlation with the actual rates is 0.9; the average percentage error is 1.9%; the squared percentage error is 0.7%; and the root mean-square error is 8.5%.

Nonetheless, there are periods when the simulated series fall short. For instance, in the late nineteenth century and early twentieth century the simulated rates are much less than the actuals. The implied β_i were too optimistic about aging then. Another way of stating it is: the implied β_i should have been smaller (or the slope of the profile much less) than what turned out by using the implied parameter. Similarly, the opposite occurs in the late 1940s and 1950s, when the implied rates turned out to be more than the actual ones.

These gaps between the implied and the actual rates very likely reflect the influence of life-course events. The implied β_i are the childhood-linked signal in the profiles of the contributing cohorts. Five cohorts contribute to the period-profile

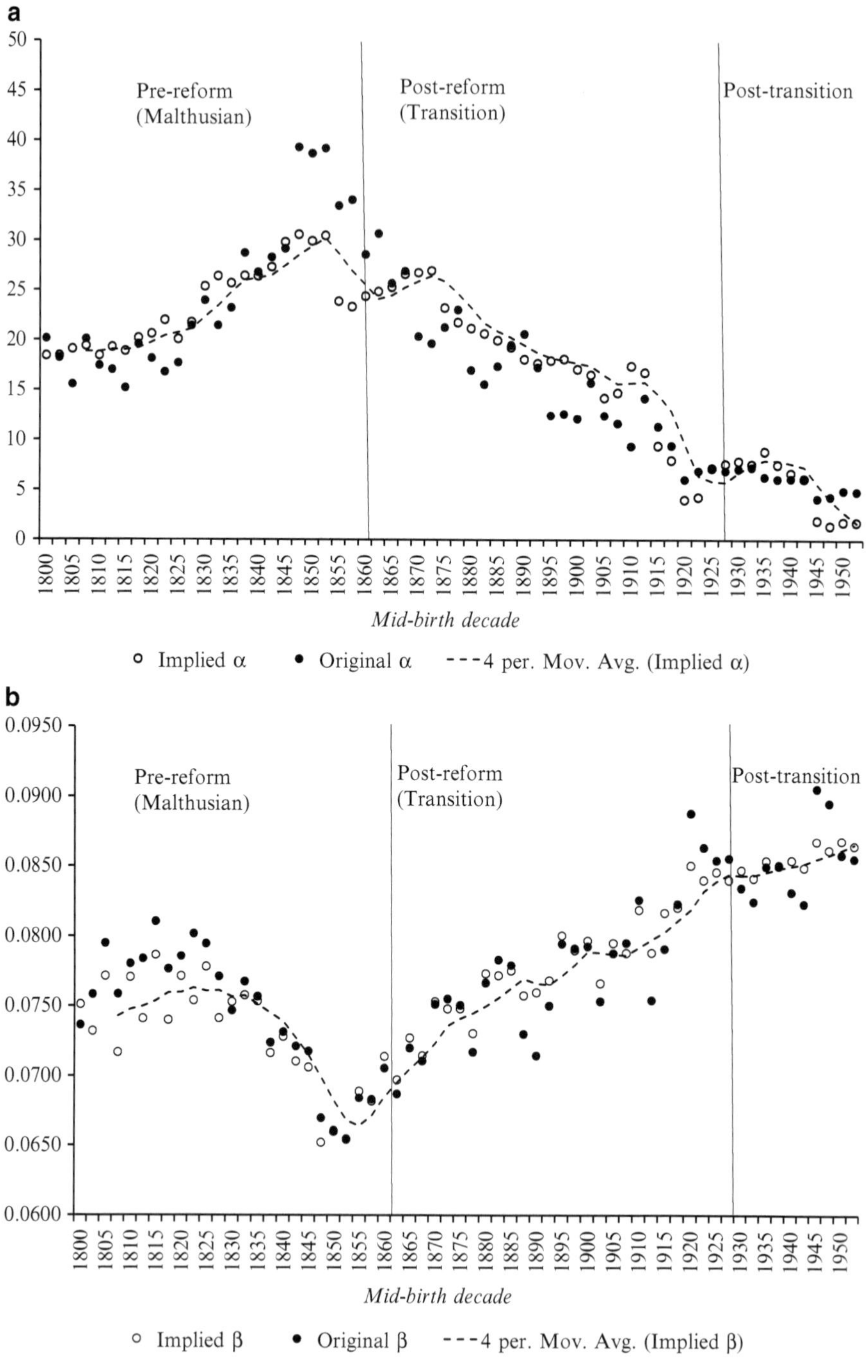

Fig. 8.4 Original cohort parameters vs. ones implied by average parameters across all childhood proxies (**a**) α_i (**b**) β_i

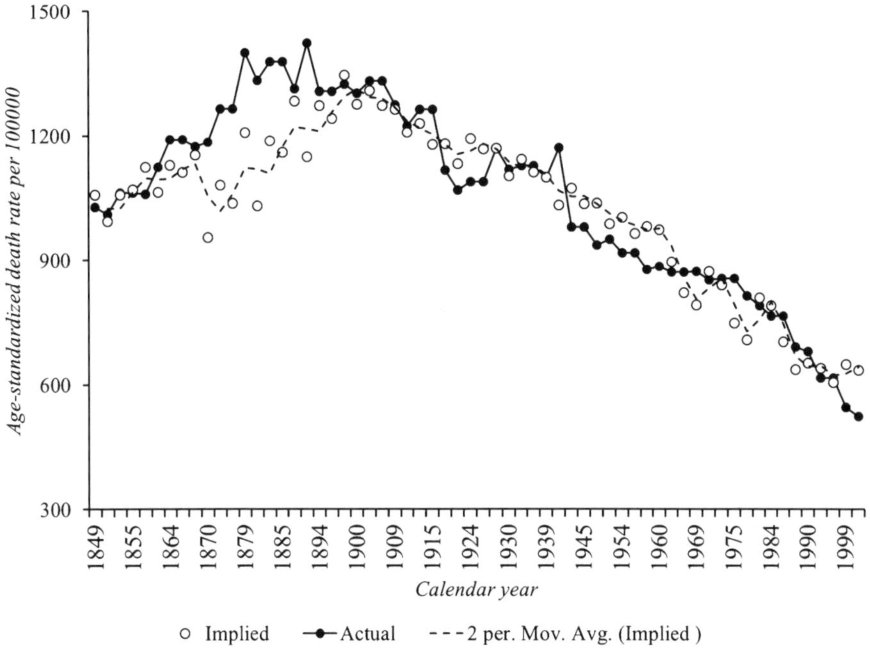

Fig. 8.5 Aggregate rates formed by implied average β_i and original α_i, vs. the actual aggregate rates (Sources: HMSO; author's calculations)

in any year. The slope of the simulated period-profile is unable to fully capture the slope of the actual period-profile. It must be that the uncaptured segment is related to life-course events, which are beyond the reach of the childhood proxies.

At this point, some notation makes things easier. Denote the slope of the actual period-profile for year t as $\beta(t)$; the childhood-linked signal in that period-profile as $\beta^C(t)$; and its life-course-related segment as $\beta^L(t)$. Assuming that the childhood-linked and the life-course-related components do not interact, $\beta(t) = \beta^C(t) + \beta^L(t)$. In the format of the study, the childhood-implied β_i in five cohorts, with each cohort at a different stage of its life cycle, helped form $\beta^C(t)$. This is akin to the simulations in Tables 4.1 and 4.2; just that the inputs are not the original parameters, but the ones implied by the childhood proxies; and, the outcome of interest is not the simulated aggregate rate, but the $\beta^C(t)$ segment of the profile.

Table 8.9 illustrates how it is calculated for four specific years. Consider the first segment for 1894. The right-most column has for various ages the actual rates as reported, with $\beta(t)$ of the 1894-profile mentioned at the bottom. The first column is the cohorts contributing at various ages to that profile. The parameters ($\widehat{\alpha}_i$ and $\widehat{\beta}_i$) are the averages of the parameters implied by the three proxies. Each cohort's parameters imply specific rates, mentioned in the column 'childhood-linked period-profile.' Thus, signals from five cohorts' childhoods make up each childhood-

Table 8.9 $\beta^C(t)$ based on average parameters across the childhood proxies

Cohort	Proxy-implied Average parameters			Age	Childhood-linked period-profile	Actual period-profile
	$\hat{\alpha}$	$\hat{\beta}$	Age		1894	1894
1860–1869	25.1	0.0727	29.5	35–44	264.5	227.3
1850–1859	23.7	0.0698	39.5	45–54	360.4	480.3
1840–1849	29.7	0.0706	49.5	55–64	978.2	967.4
1830–1839	25.5	0.0753	59.5	65–74	2,252.2	1,946.3
1820–1829	19.7	0.0778	69.5	65–74	4,393.8	4,292.6
					$\beta^C(t) = 0.0787$	$\beta(t) = 0.0728$
	$\hat{\alpha}$	$\hat{\beta}$	Age		1904	1904
1870–1879	23.1	0.0748	29.5	35–44	209.5	192.9
1860–1869	25.1	0.0727	39.5	45–54	443.7	428.0
1850–1859	23.7	0.0698	49.5	55–64	717.8	951.2
1840–1849	29.7	0.0706	59.5	65–74	1,981.7	2,048.6
1830–1839	25.5	0.0753	69.5	65–74	4,783.1	4,512.3
					$\beta^C(t) = 0.0775$	$\beta(t) = 0.0787$
	$\hat{\alpha}$	$\hat{\beta}$	*Age*		1954	1954
1920–1929	7.1	0.0846	29.5	35–44	85.6	87.5
1910–1919	9.3	0.0817	39.5	45–54	233.4	191.2
1900–1909	14.1	0.0795	49.5	55–64	721.3	530.0
1890–1899	17.8	0.0800	59.5	65–74	2,084.4	1,387.4
1880–1889	19.8	0.0775	69.5	65–74	4,328.4	3,564.4
					$\beta^C(t) = 0.1004$	$\beta(t) = 0.0940$
	$\hat{\alpha}$	$\hat{\beta}$	*Age*		1964	1964
1930–1939	8.7	0.0854	29.5	35–44	108.5	81.2
1920–1929	7.1	0.0846	39.5	45–54	199.4	193.0
1910–1919	9.3	0.0817	49.5	55–64	528.3	523.3
1900–1909	14.1	0.0795	59.5	65–74	1,597.4	1,361.6
1890–1899	17.8	0.0800	69.5	65–74	4,641.0	3,288.7
					$\beta^C(t) = 0.1064$	$\beta(t) = 0.0953$

Note: The predicted rate at each age is $\left(\hat{\alpha} \times e^{\hat{\beta}\ age}\right)$

linked period-profile, and the consequent $\beta^C(t)$ at the bottom of the column. The main contributors in 1894 were the pre-reform or the Malthusian cohorts. Similarly, in the last two segments of the Table, in 1954 and 1964, the main contributors were the transition-stage cohorts.

The approximate role of the life-course events or developments can also be seen in the segment of the aggregate rates not captured by the childhood implied rates of non-communicable diseases. Figure 8.5 indicates that life-course events may have been significant during the late nineteenth century and early twentieth century as the childhood-linked rates turned out less than the actuals. This is an example of negative life-course events or factors. They contribute by making the overall period-profile flatter (less aging), even as $\beta^C(t)$ was indicating a steeper slope (more aging). In Fig. 8.5, it is not the case that $\beta^C(t)$ was implying improvement over time. It too was indicating climbing rates. Just that the rates turned out to be worse, likely because of $\beta^L(t)$ were adverse and led to even greater rates of non-communicable diseases (more about this issue later).

A mild case in the opposite direction appears in the 1950s and 1960s. The life-course situation then was likely positive: $\beta^C(t)$ may have been indicating that the rates should be worse, but it did not turn out that way. The obvious candidate during this time is the advent of the National Health Service and the new technologies that may have become more widely accessible since.

An example may help clarify how the relative roles of the childhood- and the life-course components may change. Say there is a strong adverse link between rheumatic fever (or some infection in childhood) and deaths from heart disease sometime in the adult years (e.g. Buck and Simpson 1982; Lindholt et al. 1999; O'Connor et al. 2001). Preventing childhood infections years ago may eventually lead to lower death rates from heart diseases. This would boost $\beta^C(t)$. Now, a technology comes along, treats heart disease, and prevents deaths as well. As a life-course event, it would shrink the role of $\beta^C(t)$ and expand that of $\beta^L(t)$ in the changing $\beta(t)$ (which is not static). Note, this is not an issue of there being no $\beta^C(t)$ once the technology comes along. It is about there being relatively less of it and relatively more of $\beta^L(t)$ than in the past. Further, $\beta^C(t)$ could be exogenous to $\beta(t)$, determined by the experience of the childhood years. For instance, $\beta^C(t)$ in the second half of the nineteenth century would pertain to the childhoods in the first half of the nineteenth century. Likewise, $\beta^C(t)$ in the post-World War II era would pertain to the mid- to late-transition stage cohorts, who were the main contributors at this time.

Though averaging the parameters could have lessened some of the bias each proxy's parameters alone might have entailed, the average still has a major shortfall: if each proxy relates to specific diseases in the aggregate, then, as a matter of science it is probably better to discover those links than to worry about the statistical issue of bias which will never go away no matter how many proxies are lined up. An omitted variable, ostensibly, always exists. For instance, it is possible that infections relate to some diseases in the aggregate better than stature does. Should one worry about statistical bias and average the parameters, or find the extent of those links by using specific proxies by themselves?

Further, $\beta^C(t)$ and $\beta^L(t)$ would differ across the proxies. Their relative magnitudes would depend on when the life-course factor such as a new technology came up for the disease being studied. For instance, if a technology, or a life-course situation such as the NHS, betters the adult-age outcomes linked to past pregnancy complications more than it betters the outcomes linked to childhood infections, then $\beta^L(t)$ for the pregnancy-linked diseases would be more than the $\beta^L(t)$ for the infections-linked ones.

Table 8.10 Results of simulations with original α_i, and the childhood-implied β_i by different proxies

	Correlation	Mean E^{2*}	$\sqrt{MSE}$
Average stature	0.96	0.3%	5.7%
Infectious diseases	0.96	0.6%	7.8%
Pregnancy complications	0.96	0.5%	7.0%

Notes: N = 62 (stature); 50 (infectious disease); 44 (pregnancy complications)
*E is % error between the implied and the actuals; $\sqrt{\text{MSE}}$ is root mean square of E. Correlation is between the actual aggregate series and the simulated one

One way to see this issue is to do the same simulation, this time using the childhood-implied β_i for each individual proxy, pairing it with the original perfect α_i of each cohort. In this simulation, the β_i implied by each proxy would be the only source of error and thus of the consequent $\beta^L(t)$. Comparing the results across the proxies may then indicate the extent and when $\beta^L(t)$ varied. The summary statistics in Table 8.10 indicate that β_i implied by each proxy helps closely track the long-term trends.

Figure 8.6a–c plot the results of the simulations for each proxy. Comparing Fig. 8.6a to Fig. 8.5, it is clear that the negative life-course effects in the late nineteenth and the early twentieth century come from stature's inability to pick up the right long-term path. At least near the turning point, both infections- and pregnancy complications-linked $\beta^C(t)$ (Fig. 8.6b, c) were closer to the actuals than the ones from stature. It is plausible that if only stature were used, what appears as a negative life-course effect would become less if either infections or pregnancy variables were used instead. That is, $\beta^L(t)$ would be an adverse one (indicating a flatter slope) for stature-implied profiles, but minimal for the profiles implied by the other two proxies.

It could also mean that diseases linked to stature may have done much better, but ones linked to childhood infections and pregnancy complications did worse, and that the rates implied by these last two proxies were much closer to actuality. Likewise, for infections, the issue appears slightly later, where the rates they implied (based on the early- to- mid- transition cohorts' childhood experiences) were less than the actuals.

For the 1950s, both stature and infections-implied rates do well, but the ones implied by pregnancy complications were much higher. This is an example of a positive life-course effect. Medical care during this time may have been able to weaken the link between pregnancy complications in the past and the implied outcomes. The role of $\beta^L(t)$ in diseases linked to pregnancy complication would expand, but $\beta^L(t)$ in diseases linked to stature and infections may not.

Further, it is critical to keep in mind that diseases such as many types of cancers may defy medical therapies and (or) may not have childhood links. They would be the source of negative life-course effects for all proxies. If such diseases progressively dominate the aggregate rates, the relative role of $\beta^L(t)$ may continue to expand and that of $\beta^C(t)$ may shrink in the changing $\beta(t)$ of non-communicable diseases.

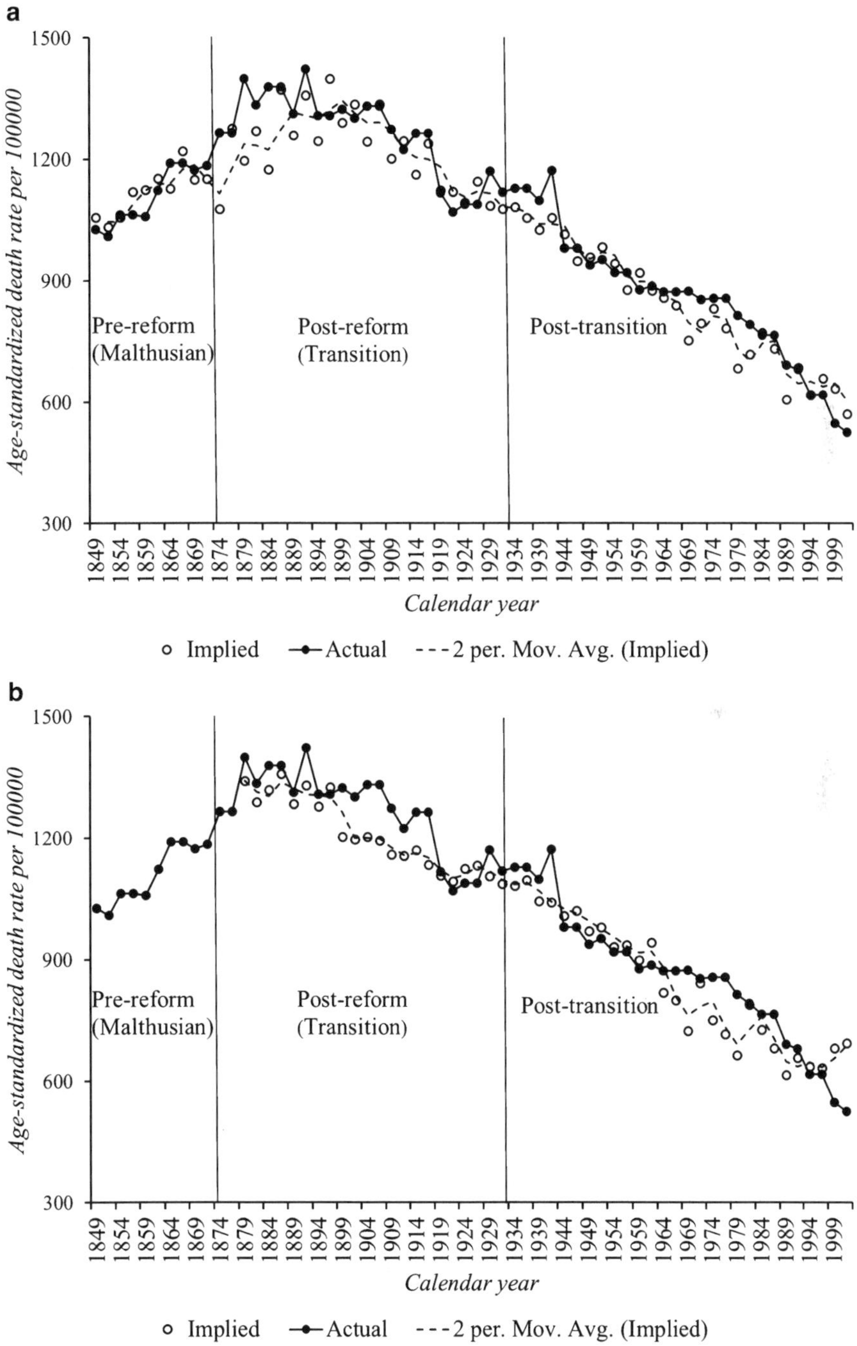

Fig. 8.6 Aggregate rates formed by original α_i and implied β_i vs. the actual aggregate rates (**a**) Average stature (**b**) Infectious disease (**c**) Pregnancy complications (Sources: HMSO; author's calculations)

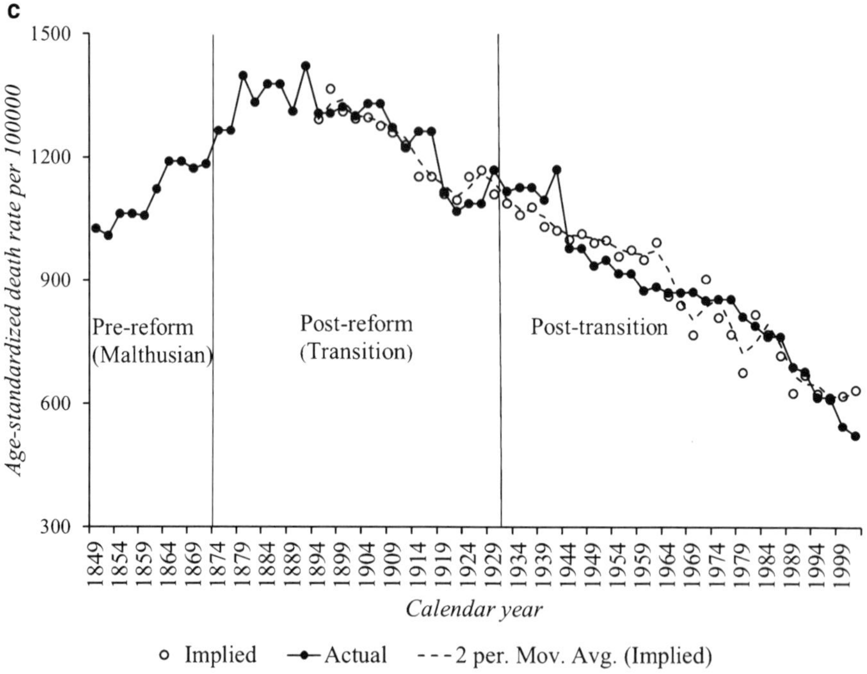

Fig. 8.6 (continued)

It is beyond the scope of the book to study specific diseases or disease sub-categories for the potential links. But this critical issue needs to be figured out. For the time being, in the next chapter, I will be using the series implied by the averaged parameter values to illustrate the role of aging in long-term healthcare spending. Chapter 10 will return to the roles of $\beta^{L}(t)$ and $\beta^{C}(t)$ in analytical frameworks of aging transitions and the potential paths of healthcare spending.

References

Angirst, J. D., & Pischke, J.-S. (2009). *Mostly harmless econometrics, an empiricist's companion*. Princeton: Princeton University Press.

Barker, D. J. P. (1992). *Fetal and infant origins of adult disease*. London: British Medical Journal.

Barker, D. J. P. (1994). *Mothers, babies and disease in later life*. London: British Medical Journal.

Barker, D. J. P. (1998). *Mothers, babies and health in later life*. Edinburgh: Churchill Livingstone.

Ben-Shlomo, Y., & Davey Smith, G. (1991). Deprivation in infancy or in adult life: Which is more important for mortality risk? *Lancet, 337*(8740), 530–534.

Ben-Shlomo, Y., & Kuh, D. (2002). A life course approach to chronic disease epidemiology: Conceptual models, empirical challenges and disciplinary perspectives. *International Journal of Epidemiology, 31*(2), 285–293.

Buck, C., & Simpson, H. (1982). Infant diarrhoea and subsequent mortality from heart disease and cancer. *Journal of Epidemiology and Community Health, 36*(1), 27–30.

Davey Smith, G., Gunnel, D., & Ben-Shlomo, Y. (2001). Life-course approaches to socio-economic differentials in cause-specific adult mortality. In D. Leon & G. Walt (Eds.), *Poverty, inequality and health: An international perspective* (pp. 88–124). New York: Oxford University Press.

Forsen, T., Eriksson, J. G., Tuomilehto, J., Teramo, K., Osmond, C., & Barker, D. J. P. (1997). Mother's weight in pregnancy and coronary heart disease in a cohort of Finnish men: Follow-up study. *British Medical Journal, 315*(7112), 837–840.

Harris, B. (2004). *The origins of the British welfare state: Social welfare in England and Wales 1800–1945*. New York: Palgrave, Macmillan.

Kuh, D., & Davey Smith, G. (2004). A life-course and adult chronic disease: An historical perspective with particular reference to coronary heart disease. In K. Diana & Y. Ben Shlomo (Eds.), *A life-course approach to chronic disease epidemiology*. Oxford: Oxford University Press.

Kuh, D., Hardy, R., Lagenberg, C., Richards, M., & Wadsworth, M. E. J. (2002). Mortality in adults aged 26–54 years related to socioeconomic conditions in childhood: Post-war birth cohort study. *British Medical Journal, 325*, 1076–1080.

Lancet. (2001). An overstretched hypothesis? *Lancet, 357*, 405.

Lindert, P. (2000). Three centuries of inequality in Britain and America. In A. B. Atkinson & B. Francois (Eds.), *Handbook of income distribution* (Vol. 1, pp. 176–316). Amsterdam: Elsevier Science.

Lindert, P. (2004). *Social spending and economic growth since the eighteenth century, volume 1 and 2*. New York: Cambridge University Press.

Lindholt, J. S., Fasting, H., Hennenberg, E. W., & Ostergaard, L. (1999). A review of chlamydia pneumoniae and atherosclerosis. *European Journal of Vascular and Endovascular Surgery, 17*(4), 283–289.

O'Connor, S., Lee, T. C., Epstein, A. C., & Libby, P. (2001). Potential infectious etiologies of atherosclerosis: A multifactorial perspective. *Emerging Infectious Diseases, 7*(5), 780–788.

Osmani, S., & Sen, A. K. (2003). The hidden penalties of gender inequality: Fetal origins of ill-health. *Economics and Human Biology, 1*(1), 105–121.

Piketty, T. (2014). *Capital in the twenty-first century*. Cambridge, MA: The Belknap Press of Harvard University Press.

Chapter 9
Aging in Healthcare Policy

9.1 A Red Herring?

Chapter 1 outlined the three main issues that are said to be vital for per capita healthcare spending in the years ahead: (1) aging, and the prospect of healthy longevity or of healthy aging for middle-to-older-age people; (2) growth of real per capita income; and (3) residual factors, conjectured to be some mix of policies, technology, and relative prices that sway productivity of the healthcare sector. The residuals are said to constitute the supply of healthcare; aging and real per capita income the demand.

Further, the contribution of residual factors in this framework hinges on an accurate assessment of the demand-side factors, for

$$g_r = g_h - \mu_y g_y - \mu_a g_a, \tag{9.1}$$

where g_h is the growth of per capita healthcare expenditures, g_y real per capita income growth, g_a growth of aging, and g_r the residual's growth. Equation 9.1 is best seen as a framework of accounting, not causality. Scholars usually estimate separately the income elasticity (μ_y) and the aging elasticity (μ_a) to roughly gauge what g_r may be. In an accounting over 1995–2005, for example, OECD (2013) uses 0.8–1.0 as μ_y, and (implicitly) 1 as μ_a, which yields g_r as the main reason for spending growth.

The finding that the residual boosts spending more than other factors has a policy implication: by containing costs, the residual can be squeezed, which, in turn, may slow spending to some sustainable rate. For example, if spending were climbing at an average annual rate of about 8%, and the residual at 3% rate, then the pace of spending can be slowed to about 5% average annual clip by reining-in factors that may be boosting the residual, especially the ones raising the inefficiency of the healthcare sector. In the European Commission's Ageing Report 2012 and in OECD (2013), for instance, this issue falls under the broader rubric of cost

S. Arora, *The Transitions of Aging*, International Perspectives on Aging 12,
DOI 10.1007/978-3-319-14403-0_9

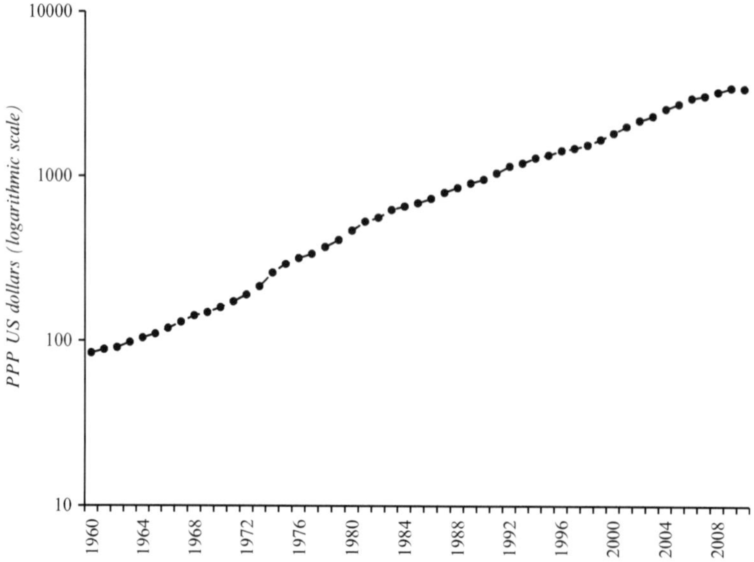

Fig. 9.1 Real per capita healthcare expenditures 1960–2010 (Source: OECD)

containment, a matter front and center in the debate on healthcare policy since the 1980s and 1990s (Schwartz 1987; Baumol 1988, 1993; Culyer 1989; Weisbrod 1991; Newhouse 1993). This chapter sheds some light on the role of aging (μ_a, g_a) in the framework, and evaluates quantitatively some of the difficult policy options that can come up at the aggregate level.

Figure 9.1 shows the annual real per capita spending over 1960–2010. Each point in this series is linked to a period-profile of aging as well as with the profile of spending. Figure 9.2 (next page) shows the spending as percentage of GDP through 2010, and its projections for the decades ahead. The numerator of this series through 2010, too, is linked to those period-profiles. The projections, meanwhile, are partly based on the outlook of aging, resting on two premises about the aging profiles: (1) they stay static; and (2) aging is signified by growing average age, which is usually measured by the dependency ratio; a bigger ratio stands for greater average age, and more aging.

By uniting the two premises, researchers have argued that aging may not play a major role in upcoming years, expressing doubt about the notion that spending is set to grow as the baby-boom generation ages ever more. The skeptics note:

> What is true for a cross-section of Americans at a point in time, however, is not a reliable guide to what happens when a country's entire population ages gradually over time. (Reinhardt 2003, p. 27)

The quote is very akin to the evidence in the book: an upward-sloping age-profile of non-communicable diseases, the cross-section of the rates of non-communicable diseases, at any point in time, does not necessarily mean an upward trending

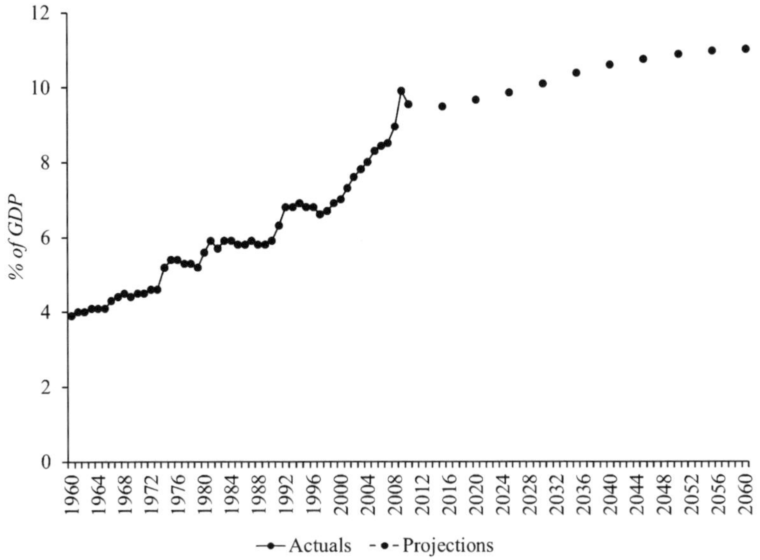

Fig. 9.2 Healthcare expenditures as percentage of GDP 1960–2010 (Sources: OECD, EU Commission Report on Ageing (2012))

aggregate rate over time. The profiles can shift down, and the trend can fall instead. It correctly states that a period-profile, the "cross-section of Americans at a point in time," cannot be relied upon as a guide to the future. Going by the evidence in the previous chapters, the reader might be expecting that a reliable guide to how the "entire population ages gradually over time" is some potential new profile. However, that does not seem to be:

> Although in any health spending for people age sixty-five or older tends to average three to five times that for younger Americans, the aging of the population is too gradual a process to rank as a major driver in health care. (Reinhardt 2003, p. 27)

The "spending for people age sixty-five or older is three to five times more that for younger Americans" admits the profile slopes up. However, the "too gradual a process" occurs along a static profile:

> To explore what effect the aging of a country's population would likely have on the demand for health care, other things being equal, one projects what the per capita use of health care or per capita health spending would be in specific future years, *if the age-specific use of health care were to remain constant at current levels over the entire forecast horizon* and only the age-composition of the population (not even the size of the population) were allowed to change in accordance with the current demographic projections. (Reinhardt 2003. p. 29, italics mine).

The main elements of this view seem to be: (1) the period-profile slopes up with age; (2) the profile's current slope is an unreliable guide to the future; (3) the future is not some new profile with a new slope, but the same static profile the population climbs as its dependency ratio gradually grows over the years; and (4) all else equal, since the ratio crawls too slowly, aging expressed by that slow-moving measure may not boost spending by much. Hence, "aging of the population is too gradual a process to rank as a major driver in health care." The emphasis on aging thus looks like a 'red herring' (Zweifel et al. 1999; Werblow et al. 2007). Breyer et al. (2011), for instance, go on to state:

> The fixation of policymakers on ageing seems to suggest that higher healthcare expenditure is inevitable.... Blaming population ageing distracts from the decisions that really ought to be made, such as devising appropriate incentives for *curbing excessive provisions of publicly financed healthcare* and evaluating the social value of new technologies. [Italics mine].

The OECD and the European Commission's Ageing Report 2012 assess the matter using a similar method. In the Report, for instance, all but one of the five main scenarios has a static profile, leading it to infer that aging will likely play a minor role (Oliveira Martins and Maisonneuve 2006; European Commission 2012a, b, Table 7.1, pp. 211; also see Cutler and Sheiner 1998).

However, the inference may be premature as the basis for it seems inconsistent. It starts with a spending profile and then alters the age-composition to study what might happen to the spending. However, changing the composition also means the profile is unlikely to stay static. The new age-composition, by definition, must have a larger fraction of old people, who may need to spend more if their rates of non-communicable diseases were greater than the rates underpinning the baseline profile. The reverse could occur if the rates for some reason were less than the baseline. The very purpose of altering the age composition, of saying that there will be more old people, is to find out what would happen if the age-profile of per capita spending, too, becomes steeper or flatter than the baseline one. This is very much like the doubt raised by Fuchs (1990) and Newhouse (1992), quoted in Chap. 1. However, in most analyses of the issue, by assumption, the outcome is read from a given stationary profile.

The issue becomes acute when the stationary profile is used to build-in the scenario of healthy longevity, in which older people are healthier than they used to be. The profile bends at middle-to-older ages, making that age-segment flatter than the rest (like the one in Chap. 1, Fig. 1.3b). It betrays an asymmetry in the underpinnings: cohorts of older people continually fare better, but the youth do not change whatsoever. Spending rates in the younger-age segment of the profile stay the same, but the slower spending built-in for the old makes the profile flatter than the baseline. Consequently, healthy-longevity or healthy aging delivers slower spending growth.

The long-term evidence on non-communicable diseases, however, convey that flatter slopes occur when a series of cohorts had been faring worse, and faring worse since early in their adult life cycles—like the series of profiles that had shifted up in the nineteenth century. For the decades-long horizon, in episodes of progress or of regress, period- and cohort-profiles had shift since early adult age (Chap. 3). Tied to the relatively flat segment of the projected profile, then, an ever-unchanging segment for the youth seems incompatible, for they too should have been doing worse, and their spending would respond accordingly.

Moreover, the projections depict the flatter parts of the profile as a good outcome, underpinned by healthy longevity, which is just the opposite of what the profiles of non-communicable diseases indicate. Could healthy longevity or healthy aging be possible in a profile of higher death rates than before? Under what situation may it accompany slower spending than the baseline? Both issues are unclear; and Chap. 10 will explore the underpinnings for such outcomes.

Meanwhile, this way of uniting a static profile and some projected measure of aging has been used for long. For instance, as part of their study on the cost of the National Health Service, Abel-Smith and Titmuss (1956, pp. 154–55) noted how strong and unrealistic the basis was for their central scenario of the cost:

> In the following exercise, therefore, we assume throughout that certain estimated population changes are the only changes which will affect the current cost of the National Health Service. Everything else remains unchanged: the incidence and character of sickness and injury; standards of diagnosis; quantity and quality of treatment; the provision of resources in goods and services; the present level of unsatisfied demand; and the present proportionate distribution of consumer use of the service by age and sex, and many other factors. There should be no need for us to emphasize the unreality of these assumptions nor the frailty of the statistical material on which we have based our calculations.

Here, the unchanged "incidence and character of sickness and injury" is the static illness profile. It was taken from the 1951 Census on the fraction of people hospitalized by age (Table 92, p. 156). The unchanged "present proportionate distribution of consumer use of the service by age and sex" is the static spending profile. The authors then scaled it up by using the Registrar-General's main population projections over 1951–1971 (Table 91, p. 155) to surmise how much the hospitalized population and the in-patient costs may grow (Table 93, p. 157), though they doubted if deaths could fall as much as the projections indicated over the horizon: "the assumption of declining mortality might be too optimistic" (p. 155).

The horizon, in retrospect, turned out to be one when the mid- to late-transition cohorts, along with new medical advances, had dragged the aggregate rates down (Chap. 4, Table 4.2). In Chap. 4, while the profiles studied in that episode of the post-World War II era were neither of illness rates nor of healthcare services, Abel-Smith and Titmuss (1956) seemed very aware of the tenuous basis for keeping the profile static, and of the possibility that it could shift in a way similar to the profiles in Table 4.2 had shifted.

They faced a very difficult practical issue: what would the new illness rate profile look like, and how would it relate to the new aging profile of the types explored in this book. Further, to come up with the costs of the NHS, they needed to find out how would the two types of profiles relate to healthcare spending.

To understand how illness and deaths relate to healthcare expenses empirically, research has over the years paid closer attention to the age and the situation in which expenses are the most. It found that expenses grow in the period leading up to death, likely reflecting the seriousness of the illness then. They also tend to be higher at older ages than at younger ones, but more so near the time of death. They have been called 'death-related costs,' associated with the proximity to death, when urgent need fosters a very inelastic demand for intense medical treatment, a combination that amplifies the expense.

Death-related costs have increasingly come to the fore in investigations of healthcare expenses. An early case was a 1984 study in the United States of the Medicare enrollees who had died in 1978; they were just 5.9% of the people in the study, but accounted for nearly 28% of the total expenses (Lubitz and Prihoda 1984). Later research found that the broad pattern had not changed much over 20 years (Hogan et al. 2001). Studies found that expenses in the year before death were large across disease-categories such as heart disease and cancer (Riley et al. 1987). A longitudinal study of last eight quarters of life over 1983–1992 uncovered that proximity to death related to expenses better than age (Zweifel et al. 1999), the growing expense sometimes visible up to ten years before death (Miller 2001). Seshamani and Gray (2004) find that both age and proximity sway hospital expenses, but proximity seems to sway more than age (also see Breyer and Felder 2006; O'Neill et al. 2000; Felder and Schmidt 2000; Werblow et al. 2007; Dormont et al. 2006).

On finding that proximity to death matters more than age, unfortunately, scholars have tended to infer that aging is likely to be less relevant than other issues in the years ahead, a view reflected in the projections by the OECD and by the European Commission's Ageing Report 2012, though such costs have been given special attention in both, especially for the long-term care segment of the total expenditures. The analytical framework in the next chapter will model such death-related costs explicitly.

Meanwhile, studies skeptical of aging, see it either as age along some static profile, or, in micro-level studies, as age, like a mere milestone, like a static trait. As an indicator of aging, average age measured from the cross-section of people of different ages, is sensitive to fertility rates: more children, all else equal, would instantaneously lower the average and signal less aging, while having little to do with the state of the people in the age-profile. Modern aging is likely neither age or average age, nor some static profile. Its main feature is the progress across profiles, and the compression such progress implies. Without the shifts, the Malthusian-type profile would be the only one left, but that is surely not the profile to use today for framing an outlook for the upcoming decades. If not, then one is forced to admit the possibility that profiles can shift substantially. If they can shift, then neither age nor a given profile may be a reliable measure of aging over the long horizon.

Before continuing, it is best to be clear about where the evidence agrees or disagrees with the skeptics. They have said that, all else equal, an upward-sloping period-profile of spending does not mean total spending will soar just because average age will grow. The evidence on non-communicable diseases strikes a very similar note: had an upward-sloping period-profile been used long ago as a guide to the future, it would have said that the aggregate rate will climb in the twentieth century when average age grew. Just the reverse ended up occurring. So the main spirit of the evidence agrees with the skeptics when they say that any given period-profile is a poor basis for framing an outlook for the longer term.

Where the evidence disagrees is the reason given for the relative unimportance of aging. They appear to use a static profile to infer that aging does not boost spending as much as others factors. However, little or no relation to aging would be found in a framework that rules out the shifts of the profiles, which is the main thrust of modern aging. Thus, the basis for being skeptical may have been weak.

9.2 No—And Not the Source of a Free Lunch Either

The question is: how may the shifting profiles as expressions of aging relate to spending? For a given finite limit to lifespan, the slope parameter of each profile, β, expresses aging: all else equal, larger β stands for a steeper slope, which reflects deaths being delayed to older ages, indicating compression, and thus more aging. This is true for both period- and cohort-profiles. I will be using the period $\beta(t)$ as the measure of aging, and then later bring in its childhood-linked and life-course components from Chap. 8.

Figure 9.3 shows the annual growth of real per capita healthcare spending (right axis) along with the growth in the $\beta(t)$ of annual period-profiles (left axis), both 4-year moving average. Denote growth of $\beta(t)$ as $g_\beta(t)$.

In the 1960s and 1970s— the early stages of the age of delayed deaths—$g_\beta(t)$ had sped up. Non-communicable diseases in middle-to-older age people were falling rapidly, and would later prompt the modern debate. The period-profiles were turning steeper. However, growth of aging subsequently slowed, turning into shrinkage since mid-to-late-1980s. This occurred, ironically, just when the debate about aging, especially the one based on micro-level data, was pointing out that delays may have become a lasting new feature.

For that reason alone, the adverse turn since the mid-1980s needs to be questioned: are the $\beta(t)$s telling an accurate story about aging? The oldest age in the profiles is 74. Recall from Chap. 3 that age 27–74 was selected to stay consistent over the 150 years of data because the reporting format had changed many times over the decades. If the delays had intensified since the 1960s, then more people would have surpassed 74 years of age. Not including them may misgauge the true $\beta(t)$. Thus, a check: for this period, calculate $\beta(t)$ for the profiles including the 75–84-year olds and check if they differ from the ones in Fig. 9.3. Beyond that age, the data during this time are reported as 85+ and not assigned an age-interval. Consequently, if a disproportionately large fraction of people had been surviving past even 84 years of age, the results could remain erroneous.

In Fig. 9.4, the solid line is $\beta(t)$ from Fig. 9.3, and the other one is the new $\beta(t)$ for profiles that include 75–84-year olds. The two are nearly identical, with correlation 0.94, although since the 1990s shrinkage in the new $\beta(t)$ is slightly less even as both series have moved in tandem. For analysis, it implies that results based on data in

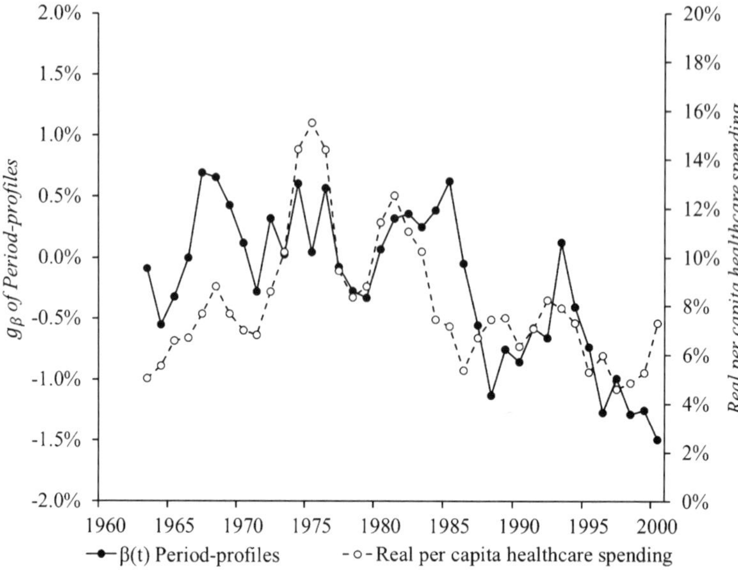

Fig. 9.3 Growth of $\beta(t)$ and of real per capita healthcare spending (4-year moving averages) (Sources: OECD; author's calculations)

Fig. 9.3 may not be far from actuality, and may closely approximate aging among 25–84 year-olds during this time.

For the debate on aging it implies two other critical things: (1) technically, by age 74, the parameters likely capture most of the vital features of the profiles since early adult age; (2) the age of delayed deaths likely waned soon after scholars had discovered it; the more inclusive $\beta(t)$s in Fig. 9.4, too, have been falling since the mid-1980s. Chapter 10 will shed some more light on this issue in terms of the childhood-linked and life-course-related components.

Meanwhile, Fig. 9.3 also clarifies that a static profile could be incapable of properly relating aging to growth of spending $g_h(t)$. Such a profile's $\beta(t)$ is constant, its $g_\beta(t) = 0$, and it would help find very little—if any—influence of aging. For reasons unclear, for instance, OECD (2013) implicitly assumes $\mu_a = 1$ even as the profile it uses is static. While $g_\beta(t)$ in that framework is zero, the measure of aging is the dependency ratio, which hardly fluctuates annually, sometimes even over several years. It is also unclear how the dependency ratio automatically relates to the underlying demand for healthcare, especially over the adult-age segment. In contrast, an upward-sloping profile for non-communicable diseases directly relates to such demand, at least in the sense that along any such profile very serious illness must have occurred for people to have died.

Moreover, a flatter profile may have been an incorrect depiction of healthy aging. As a result, ramification for spending could have been misread. Healthy aging has been linked to slower spending. However, Fig. 9.3 indicates that growing

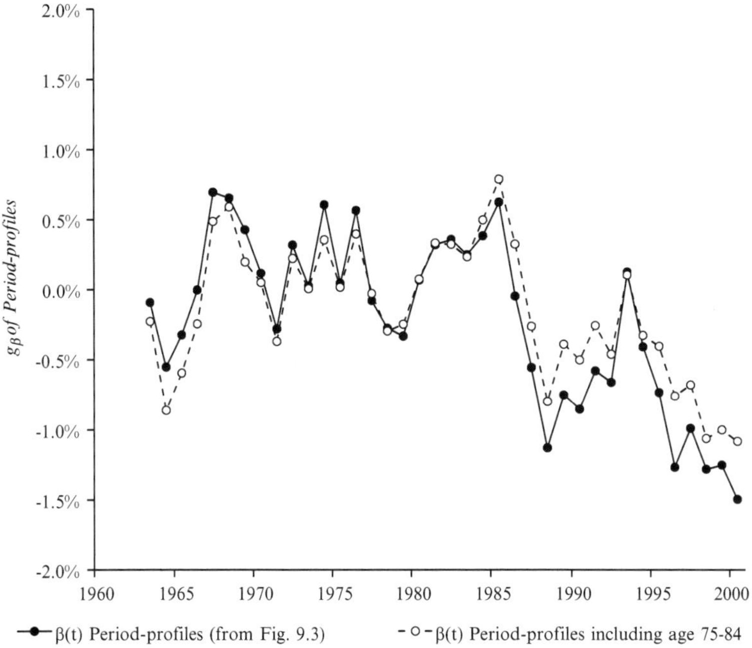

Fig. 9.4 Growth of $\beta(t)$s (period-profiles, 4-year moving average)

$\beta(t)$s, a series of steeper profiles, and more aging, coincide with faster spending. Shrinking $\beta(t)$s, a series of flatter profiles and less aging, associate with slower spending.

The question is which—steeper profile (more aging) or flatter profile (less aging)—is accompanied by healthy aging, the one with less illness rates than before? Say the flatter profile goes with healthy aging. One must then face the fact that such profiles have higher death rates at each age. What is 'healthy' about people dying at greater rates since early adult age? Could the illness rates have been falling? If not, why is this situation healthy even though spending is slower? Now say instead that a steeper profile accompanies healthy aging. The evidence then says that it goes with faster, not slower, spending. Which type of aging profile is accompanied by healthy aging is thus an empirical issue about the type of illness profiles (and their $\beta(t)$s). The next chapter explores them, and discusses what type of spending profile may be observed in various combinations of illness and aging profiles.

Meanwhile, say the relation of per capita healthcare spending $h(t)$ and aging $\beta(t)$ can be expressed as

$$h(t) = \beta(t)^{\mu_a}, \tag{9.2}$$

where μ_a is a parameter. Taking natural logs of both sides yields

$$\ln h(t) = \mu_a \ln \beta(t). \tag{9.3}$$

Figure 9.1 strongly recommends their relation would be best estimated in a dynamic framework, including a time trend, such as

$$\ln h(t) = \pi_1 + \gamma_2 t + \mu_a \ln \beta(t) + \Gamma + \epsilon(t) , \tag{9.4}$$

where ln $h(t)$ is the natural log of real per capita healthcare spending; π_1 is a constant; t is a time trend; $\Gamma = \sum_{i=j}^{-j} \Delta \ln \beta(t-i)$ are lead and lag terms; and $\epsilon(t)$ is an error term. Here, μ_a is the coefficient of interest, the aging elasticity of per capita spending. It is estimated by performing Dynamic Ordinary Least Squares equipped with the Newey-West method (with lags = 2) that corrects for heteroskedasticity and serial correlation, and for $j = 1$. The outcome for the available 38 observations, and 32 degrees of freedom, is:

$$\begin{aligned} \ln h(t) = \;& 12.3 + 0.087t + 3.44 \ln \beta(t) \\ & (21.6) \;\; (49.7) \quad\;\; (13.8), \end{aligned} \tag{9.5}$$

$$\overline{R}^2 = 99.7\,\%$$

where t-statistics are in parentheses, and estimated errors $\hat{\epsilon}(t)$ were tested to be white noise. (I will be using central values of the estimates and average of the annual growth rates for the discussion below). The estimates indicate that an average annual 1% increase in $\beta(t)$ boosts per capita healthcare spending by about 3.4%.

In the post-World War II era, there appear to have been two periods of aging: the age of delayed deaths or the compression over 1945–1984, when $\beta(t)$ grew at an average annual rate of about 0.63%; and then 1985–2000, when it shrank at 0.82% average annual clip. According to OECD data, average annual growth of $h(t)$ or g_h through 1984 was about 8.8%; after then, in the era of cost containment, it slowed to about 6.5% average annual rate. All else equal, over that first segment, aging boosted $h(t)$ by about 0.63% × 3.44, or by roughly 2.2 percentage points, accounting for about 25% (2.2 ÷ 8.8) of its growth. In the latter segment, it helped shrink spending by about 0.82% × 3.44 or 2.82 percentage points, or the entire slowdown in g_h from the earlier period.

Say income elasticity μ_y is 1, which means that per capita spending grows in proportion to real per capita income growth. The history of long-term real per capita income growth g_y is about 2%. Use 3.44 as μ_a, and use 0.63% and −0.82% as g_a over the two segments when average annual g_h was 8.8% and 6.5%. Insert the values for 1960–1984 in the accounting Eq. 9.1:

$$g_r = 0.088 - 1.0 \times 0.02 - 3.44 \times 0.0063 = 0.046, \tag{9.6}$$

which is shown as the middle outcome (of the last column) in the top half of Table 9.1. There are other potential outcomes, depending on the three values of μ_y.

Table 9.1 Accounting for growth of per capita healthcare spending, 1960–2000

	g_h	$\mu_y g_y$	$\mu_a g_a$	g_r
1960–1984				
$\mu_y = 0.8$	8.8%	1.6%	2.2%	5.5%
$\mu_y = 1.0$	8.8%	2.0%	2.2%	4.6%
$\mu_y = 1.5$	8.8%	3.0%	2.2%	3.6%
1985–2000				
$\mu_y = 0.8$	6.5%	1.6%	−2.8%	7.7%
$\mu_y = 1.0$	6.5%	2.0%	−2.8%	7.3%
$\mu_y = 1.5$	6.5%	3.0%	−2.8%	6.3%

Notice first a larger μ_y shrinks the role of the residual. An elasticity of more than 1 means that demand grows more than proportionately than real per capita income. All else equal, healthcare demand expands more, but it happens at the expense of the heft of the supply side or of the residual. The shrunken values of g_r may suggest that contribution of the supply side has been less important, but only in some mysterious 'residual' way. Since its constituents are not known with precision, what is gained or lost as one switches from one row to the other is hard to judge.

For all scenarios in Table 9.1, the more relevant takeaway should instead be that not accounting for aging properly places an undue emphasis on the residual, and on the policies that seek to contain costs embodied in the residual. What it also means is that cost containment may not be a free lunch. Something else in the accounting must give. Set aside for now the issue of what is the right value of μ_y (to be discussed in the next section). The top section of Table 9.1 says that reining-in healthcare must entail cutting resources from the very activity that could have been shrinking non-communicable disease and growing the $\beta(t)$s. Cost containment could then hinder aging and wellbeing. It helps to keep in mind that aging is defined as the shift in the entire profile, and not just the state of middle-to-older age people, which is the main focus of micro-level studies.

For the projections in OECD (2013) and European Commission (2012), or in any framework that keeps the profile static, this type of difficult choice does not surface clearly. Since the profiles do not shift, aging does not play a major role in the arithmetic to begin with. In turn, that leaves an impression of cost containment being a harmless attainable goal, at least analytically.

In Table 9.1, in contrast, aging matters to spending growth. In the bottom half of the Table, for instance, as g_a shrank, g_h had slowed. It is very important to be clear about what it conveys. It does not say that slower spending had caused the adverse outcome. It could well be the other way around. The next chapter relates this shrinking g_a to the outcomes for the inter-war cohorts, among other issues.

Meanwhile, it is quite possible, perhaps undeniable, that the healthcare sector has been productive. Its effort has saved lives, helped the profiles tilt steeper, and boosted $\beta(t)$. Some of what is being attributed to aging is actually the contribution of the sector. The accounting does not reveal it that way because it calculates the sector's role as a residual matter. How much the sector may have been contributing to aging, then, ought to be isolated as much as possible. It would make the choices more explicit than they are in the form of the residual.

How much of the growth in $\beta(t)$ can be attributed to $h(t)$? Both $h(t)$ and $\beta(t)$ can sway each other. More spending, say, on new medical technology, may shrink death rates from many non-communicable diseases, making period-profiles steeper and lifting $\beta(t)$.

One could also easily reckon that many such diseases are exogenous (e.g. cancers), even involuntary (e.g. cancer, stroke, cardiac arrest), and people usually do not have prior knowledge about which disease is going to strike them and when. Urgent medical attention then fosters an inelastic demand, often at great expense, increasing $h(t)$. However, for a given supply of medical treatments, a series of steeper profiles that had shifted down, each imbued with fairly inelastic demand, may result in less healthcare spending than before. That is, downward shifts in the profiles could coincide with more 'delayed' deaths, slowing demand. But that would occur only if illnesses were proportionately delayed too—or illness profiles had shifted down.

However, if for a given supply of treatments, illness rates had not budged or had grown instead, then steeper aging profiles (or growing $\beta(t)$) may coincide with more spending. This would be the case of static profile of illness, or even of expansion of illness (illness-profiles shifting up), coinciding with compression of deaths; and thus of falling fatality ratios (death rates divided by illness rates), but not necessarily slower spending than before.

It is possible that aging profiles closely relate to the issue of death-related costs discussed in the last section. All else equal, an inelastic demand near the time of death combined with the delayed deaths across profiles may lower death-related costs, in real absolute terms, and as a fraction of total spending. The opposite may occur if the profiles shift upward and illness rates grow. After people have discovered their non-communicable illness (i.e. information about the state becomes available), many such diseases tend to be chronic (e.g. arthritis), prone to relapse (e.g. cancers), requiring continual or repeated medical attention (e.g. chemotherapy), entailing recurring expenses before death.

The critical technical issue in all of this is: to what extent can $\beta(t)$ be exogenous—or be seen as exogenous—to current healthcare spending? If there is a way for it to be exogenous, then how may it shed light on the policy options?

The results from the previous chapter may offer a way. It had asked: the parameters of cohort-profiles were swayed by which factors in the childhood years? They were stature, rates of infectious diseases, and pregnancy complications in the decades in which cohorts were born, each along with the functional distribution of national income in the birth-decades, all of them occurring as the polity was transformed from one controlled by the elite to one increasingly set by the franchised masses.

Those factors had implied childhood-linked α and β for the cohorts. The two parameters had then generated age-specific rates in the cohort profiles. The rates in five running cohorts combined, in turn, had helped simulate childhood-linked segment of the period-profiles. The main contributors to the period-profiles after

Table 9.2 Accounting and the components of aging

	g_h	$\mu_y g_y$	$\mu_a g_\beta^C$	$\mu_a g_\beta^L$	g_r
1960–1984					
$\mu_y = 0.8$	8.8%	1.6%	0.8%	1.4%	5.0%
$\mu_y = 1.0$	8.8%	2.0%	0.8%	1.4%	4.6%
$\mu_y = 1.5$	8.8%	3.0%	0.8%	1.4%	3.6%
1985–2000					
$\mu_y = 0.8$	6.5%	1.6%	−2.6%	−0.2%	7.7%
$\mu_y = 1.0$	6.5%	2.0%	−2.6%	−0.2%	7.3%
$\mu_y = 1.5$	6.5%	3.0%	−2.6%	−0.2%	6.3%

World War II were mid-to-late transition stage cohorts. Their childhoods were in the 1880s through the 1930s. There does not seem to be a reason for spending in the post-World War II era to have influenced those years and caused those parameters; they were based on what cohorts had faced decades earlier as children.

It is very likely, then, childhood-linked segment of the period-profiles in the post-World War II era—the $\beta^C(t)$ in the last chapter—were technically exogenous to spending during this time. Denote its growth as $g_\beta^C(t)$. The other component, the one susceptible to current events, is the life-course linked $\beta^L(t)$; denote its growth as $g_\beta^L(t)$. Assuming no interaction between them:

$$g_\beta = g_\beta^C + g_\beta^L . \tag{9.7}$$

Aging's contribution is $\mu_a \times g_\beta$; multiply both sides by μ_y:

$$\mu_a g_\beta = \mu_a g_\beta^C + \mu_a g_\beta^L . \tag{9.8}$$

Over 1945–1984, average annual $g_\beta^C(t)$ was 0.23% and $g_\beta^L(t)$ 0.4%; over 1985–2000, $g_\beta^C(t)$ was −0.76%, and $g_\beta^L(t)$ −0.06%.

In the top half of Table 9.2, of the 2.2% contribution of aging, 0.8% came from $g_\beta^C(t)$ and 1.4% from g_β^L. Thus, childhood-linked aging may have accounted for about 36% of overall contribution of aging to healthcare spending. The rest could either be a cause or a consequence of healthcare spending, depending on the extent to which the underpinnings of g_β^L can be said to be exogenous to spending.

Say now that policy seeks spending at some 'sustainable' rate, defined as a rate closer to the long-term real per capita income growth (about 2%). The policy thus wants to trim g_h from 8.8% to 3%, a 5.8 percentage point cut over the long term. What are the options at hand? Constricting income elasticity is likely outside the policy's purview (discussed further in the next section). Raising the rate of long-term real per capita income growth is also difficult, certainly beyond a policy that seeks slower growth of the healthcare sector. Meanwhile, it is unlikely that $g_\beta^C(t)$ can be changed, for it is a lasting ramification of the past, of the momentum of history.

The policy may then only target the residual. However, by definition, one really does not know clearly what is in it and how it may relate to aging. It may well be that the residual is the expense incurred for growing the life-course segment of aging, assuming that *all* of the life-course changes in aging were endogenous or were caused by healthcare spending embodied in the residual. That is, g_β^L is the contribution of the healthcare sector's supply side, and they relate positively.

Squeezing the residual's growth could then involve hurting the growth of aging. If one uses any row on the top half of Table 9.2 as an example of the choices at hand, the 5.8% cut in g_r would not just get rid of the residual, it would also nearly eliminate g_β^L. The assumption is that expenses in the residual were not already involved in the 1.4% g_β^L. If they were, g_β^L may shrink just when the implementation of such a policy begins.

Say now the policymaker knows precisely what is in the residual and wants to neatly slice out only its costlier or the 'inefficient' parts while boosting the quality of care. The options it faces are just as bad. In a situation when both g_r and g_β^L are growing, smaller the available g_r and larger the g_β^L it can affect, greater the likelihood of adverse consequence.

Now consider the situation since the mid-1980s in the bottom half of Table 9.2. Here, $g_\beta^C(t)$ is -2.6% and $g_\beta^L(t)$ -0.2%. The profiles are already worsening, more because of poor $g_\beta^C(t)$ likely coming from the cohorts born in the inter-war years, or from the possibility that many diseases in the profiles have been defying medical treatment, or that the treatments do not exist. In this situation, the policymaker really does not have much of a choice. Slowing g_h from 6.5% to 3% seems to give an impression that the residual may be squeezed safely. However, if the resources already devoted to the life-course component are not enough to help it grow, then slashing them could make things worse.

Now relax the assumption that $\beta^C(t)$ and $\beta^L(t)$ grew independently. What can their interaction mean? Consider an example—say, stature and stents. One is a childhood variable, the other a life-course medical advance that came up relatively recently. The interaction says that some of what is being attributed to stature (and to economic inequality in childhood) is actually stature plus the good luck of stents having become available during the life-course of some cohorts. Stature can take you only so far; you also need stents to age more.

The consequence is that it would take some of the spotlight away from both $\beta^C(t)$ and $\beta^L(t)$, and shed it on an entity $\beta^{CL}(t)$ and its growth g_β^{CL}. It would be the third component of g_β in Table 9.2. To pick a number, say it is on average 0.4 percentage points. It trims g_β^C from 0.8% to 0.6%, and shears g_β^L from 1.4% to 1.2%, so that the three sum as $0.6\% + 0.4\% + 1.2\% = 2.2\%$.

Does it change the available options? It only says that childhood-linked aging is less, one that policy cannot reverse. More room in the other two, however, does not mean that trimming them provides a better set of options than earlier. In fact, it makes the choices harder because the roles of the life-course components (g_β^{CL} plus g_β^L), the ones more susceptible to being damaged, have expanded. So, relaxing the assumption of no interaction may add some insight, but may not offer better policy choices. (All of these results will be questioned in the next chapter).

Discussing the types of options for cost containment, Culyer (1989), for instance, pointed out "it is hard to resist the conclusion that the selective use of instruments that appear to bear on these components [finance, provision, and regulation] offer the currently best way forward. But it should always be borne in mind that cost-containment in itself is not a sensible objective. The ultimate objective of any system of health care is to promote the health and welfare of its clients,. . . " (Culyer 1989, p. 37). The discussion in this section makes explicit some of the difficulties involved when aging can be affected by those components. The next chapter clarifies further the instruments that could be availed.

9.3 What Is the Role of 'Income Elasticity'?

The other piece in the accounting is income elasticity μ_y. As the share of the health sector grew in the national incomes of OECD countries over the post-World War II era, one of the likelier reasons offered for its growth was advancing real per capita incomes, and that μ_y is more than 1. To see how, say μ_y were 1.5. Every 1% rise in real per capita income boosts spending on healthcare by 1.5%. As that extra 0.5% compounds over decades of real income growth, the sector's share in the economy expands. The share would shrink if μ_y were less than 1, and stay the same if it were 1.

Table 9.3 shows the elasticity estimated by various studies at the micro-regional- and the aggregate-level (Getzen 2000). Nearly all aggregate-level μ_y are more than 1, most micro- or regional-level are less than 1. Which one should be selected for analysis?

Neither does the wide range of the estimated elasticity mean that using any of them is appropriate, nor is the accounting a trivial exercise of generating results that can be interpreted any which way. Which μ_y is selected, and how it is construed, is a very critical issue with serious consequences for wellbeing. For example, Blomqvist and Carter (1997) quote *The Economist*, who stated as a matter of some conventional wisdom, that "as with luxury goods, health spending tends to rise disproportionately as countries become richer." What it means is that μ_y is more than 1. But luxury is something you can do without, like that latest Lamborghini, or that fourth Ferrari, you could forego. It is the wrong connotation and a highly misleading way of conveying what is at stake.

To see it better, consider Hall and Jones (2007), where healthcare demand is an expression of the desire to extend life. It lets people escape the waning marginal utility from material goods. Such goods, though vital, can get you only so much fulfilment: having one or two cars may be useful and enjoyable, say, but a third car may not add as much to either of those experiences as the first two. Further, the utility each car offers is as short-lived as the car itself. Whereas, living healthier and longer trumps that waning satisfaction: it lets you enjoy any number of cars, even as

Table 9.3 Studies of income elasticity of healthcare spending

	Income elasticity (μ_y)
Micro-level	
Newhouse and Phelps (1976)	≤0.1
Manning et al. (1987) (Rand)	≈0
Wagstaff et al. (1991)	≤0.1
American Medical Association (1978)	1.0–1.7
Anderson and Benham (1970) (physician expenses)	0.4
Feldstein (1973) (dental expenses)	1.2
Parker and Wong (1997) (total expenses)	0.9–1.6
Regional-level	
Feldstein (1971) (47 states, hospital expenses)	0.5
Fuchs and Kramer (1972) (33 states, physician expenses)	0.9
Levit (1982) (50 states 1966, 1978, total expenses)	0.7
Di Matteo and Di Matteo (1998) (Canada, provinces, 1965–1991)	0.8
Aggregate-level	
Abel-Smith (1967) (33 countries, 1961)	1.3
Kleiman (1974) (16 countries, 1968)	1.2
Newhouse (1977) (13 countries, 1972)	1.3
Maxwell (1981) (10 countries, 1975)	1.4
Getzen (1990) (US, 1966–87)	1.6
Scheiber (1990) (7 countries, 1960–87)	1.2
Gerdtham et al. (1992) (19 countries, 1987)	1.2
Getzen and Poullier (1992) (19 countries, 1965–86)	1.4
Fogel (1999) (US long-run)	1.6

Sources: Selected studies from Getzen (2000); also see OECD (2006)

the enjoyment each one brings wanes. For that reason, people set aside a disproportionate share of income to healthcare: μ_y must be more than 1. The connotation, note, is way different from a luxury that can be done without; it is instead a necessity for enjoying everything else life may bring.

The reasoning in Hall and Jones (2007) is shored up by very high estimates scholars find of the value people place on health and life (Nordhaus 2005; Murphy and Topel 2003). Though one may question the methods and the assumptions, the more vital point they drive home is that policy should take into account high level of importance people attach to the issue. If they place great value, and if there is technology, then the growing share of the sector must also be socially optimal. If so, policy should be leery of subduing its growth. It is easy to see the issue in the top half of Table 9.1: where μ_y is 1.5 or 1, trimming the aging component, even as a side effect of reining-in the residual, can mean that policy is at odds with what people may value immensely.

Nonetheless, this long-term view needs to be reconciled with recent micro- and regional-level studies that find μ_y less than 1 (e.g. Acemoglu et al. 2013). The smaller micro- and regional-level μ_y imply at least four inter-related things: (1) the role of real per capita income growth alone is not that big; (2) healthcare spending may grow less than proportionately as real per capita incomes grow in the years ahead; (3) whether the sector's size shrinks as a percentage of GDP will depend on issues other than real per capita income; and (4) the sector in the past grew for key reasons that may not have been grasped well.

For the sources of spending growth other than income, the obvious one is the advent of the NHS. Before then, it was likely the National Insurance Scheme that widened access to medical care, and going further back, the public health reforms in the late nineteenth century that boosted spending on public health infrastructure. All of them have strong political roots. The chapter will come back to this issue, but for now consider what it implies for the selection of the right μ_y.

The purpose of universal health insurance is that individual incomes should not matter so much for medical needs (just as one of the main results of public health reforms several decades earlier was to even out as much as possible the epidemiologic situation across income groups). If you are rich or poor, you have access to medical care; if you become richer or poorer, you have access to medical care; if economic distress strikes people unsymmetrically, risk-pooling says they have access to medical care. If individual incomes should not matter as much under the NHS, would not the size of the micro-level μ_y be more or less irrelevant to the very long-term issue at hand? This could be one reason for not picking the short-term micro-level μ_y in Table 9.3.

Another possibility is that the horizon of most micro- and region-level studies is too short compared to decades over which aging and the accompanying demand for healthcare unfolds. Say each period-profile of non-communicable diseases represents a vital segment of the overall demand. It is made up of demands by many cohorts, each one with its own illnesses and income levels. Suppose now that income rises and the period-profile shifts. The question for estimating elasticity is: how much is the change of income (and whose income) and what is the change of demand (and whose demand)? Even if short-term shifts betray inelastic demand (summed up across the cohorts in the profile), the relevant questions are: (1) what is it over several decades of such shifts; and (2) is it invariant for the purpose of framing an outlook decades out.

At another level, cohorts age over decades. Each has its own profile of illness and incomes. How has demand evolved across them as their income profiles varied? Which cohorts would be shaping the future period-profiles of spending, and what μ_y should be used for projecting their demand? It is unclear if μ_y in short-term studies and randomized experiments—usually with poor external validity—can be useful for scenarios that play out over decades. This could be the second reason for not selecting the short-term micro-level μ_y.

The third reason is that small micro-level μ_y may be unsuitable for aggregate-level setups that can have their own μ_y (Getzen 2000). Both can be correct, just not substitutable. In the UK, for instance, aggregate-level μ_y may reflect policy change and budget allocations set at least partially by voter preferences for the NHS. Income elasticity built-up from the micro-level—if relevant under universal insurance at all—may not reflect the issue. If so, instead of studying response of healthcare demand to individual incomes, the proper design may be to study the response of aggregate-level budget allocations to systemic economic setbacks and to political change. Income elasticity, then, may be a misnomer, too.

For that very reason μ_y might not be a very useful parameter for long-term analyses or projections. They would require framing an outlook for future systemic dislocations and political changes. By studying such events in the past, any number of scenarios could be simulated for sure. But could they be the baseline scenario of the future path of spending over several decades? Could one foretell, for instance, which political party is going to administer at which time over the next few decades, and what type of systemic economic dislocations could be in the offing? If such contingencies cannot be the baseline scenario then the debate on the size of μ_y might be more about grasping the past instead of about informing an outlook. One cannot be so sure that "it [income elasticity] is necessary for forecasting how health care spending is likely to evolve in the coming years" (Acemoglu et al. 2013).

Further, it is unclear if knowing μ_y really clarifies the policy options. Consider the choices in Tables 9.1 and 9.2 and ask: should the policymaker target μ_y or the residual? Say the "true" "deep-rooted" and invariant μ_y is 1.5, the third scenario (in Tables 9.1 and 9.2), but you really do not know it, or do not have a good grasp of its size. Say also the policy goal is to slow spending to some sustainable rate. You know precisely the inefficiencies that can be removed, and advocate neatly slicing them out ensuring that aging goes unscathed. The broader effect of doing such a thing is equivalent to saying that the policymaker should rein-in particular aspects of healthcare spending, and keep it about there for a given path of real per capita income. Prescribing that the residual be cut in a particular way is then effectively an aggregate-level prescription for a smaller μ_y, something like the first or the second row in Tables 9.1 and 9.2, away from its "true" but unknown value. In this case, μ_y comes up as an outcome, not as the driver of spending; knowing its "true" value in advance may thus not have mattered.

Now say instead that one really knows μ_y, but does not know what is in the residual. The policy proposal is that spending should be slowed by adopting the new μ_y less than 1, making spending less sensitive to a given path of real per capita GDP. This could potentially do a lot of harm for one cannot tell where the impact is going to fall.

It seems that if one does not know μ_y but knows in detail what can be cut in the residual, one could end up with any μ_y, but have a better chance of preventing an adverse spillover into the other components of healthcare spending such as aging. The question here is: which one can be grasped better for practical policy purposes? It is unclear if knowing a "deep-rooted" invariant μ_y—if there is one—has more practical value than knowing the sources of inefficiency, or the quality of factors that make up the residual (Culter 2014).

Moreover, say politics and budgetary allocations do implicitly set some μ_y by deciding how responsive the spending is going to be for any given path of aggregate income. Then the estimates of μ_y in micro- regional- and aggregate-level studies may not be that useful for the purpose of projections decades out. What the policy has done in this instance is obviously not the equivalent of the usual income elasticity, but some aggregate-level "income elasticity" nonetheless. The policymaker at each point in time can choose how to respond to the state of the polity, to rates of economic growth, to systemic dislocations, to underlying demands of aging, and to the technological options at hand. In which case, it is unclear why would one need to estimate and adopt some invariant μ_y for framing an outlook.

The discussion above has four seemingly-awkward elements: (1) very high value people place on health and aging; (2) Hall and Jones (2007) assertion that micro- and aggregate-level μ_y is more than 1; (3) the likely irrelevance of micro-level μ_y under a system of universal health insurance such as the NHS; (4) the potential irrelevance of estimating aggregate-level μ_y if politics govern spending, and if future politics and systemic dislocations are unforeseeable. How do they fit in?

One may say that people strongly value health. That is valid by itself and can stand on its own. But it does not mean that income growth is what causes people to value health and longevity highly. Income elasticity, large or small, may be unnecessary for making the case that people place much emphasis on their health and longevity.

Suppose one says no, it is necessary, the two are intertwined. Consider the nineteenth century, when non-communicable diseases were climbing for four-five decades even as per capita real incomes were growing. One could say that people just did not allocate enough to better their situation despite rising per capita incomes. But that undercuts the very premise that people allocate more than proportionately as incomes grow. Or, you could say that people did not value their lives enough back then to bother allocating for healthcare out of growing incomes. It undermines the other premise that people highly value their lives and health. Both arguments do not fare well in that episode. The only way out seems to be that people did value health immensely, but were unable to allocate or did not allocate disproportionately for reasons other than income growth. What could those reasons be?

One may say two things: (1) despite income growth there were no technological solutions or effective therapies for most non-communicable diseases then, and hence no reliable or attractive avenues to spend growing amounts to remedy the situation; (2) for the first three-quarters of the century, despite per capita income growth, the franchise had not been extended enough to allocate to public health spending, and that children's state was so appalling that it took a toll in terms of rising rates of non-communicable diseases for which there were no effective medical therapies then. If one agrees with either or both, one must admit the possibility that institutional change and technology, not income growth, per se, may be the more significant drivers of the allocations to healthcare spending (Arrow 1963; Akerlof 1970; Newhouse 1992, 1993; Acemoglu and Robinson 2000).

As the case for many types of cancers illustrate even today, it should be obvious that without effective medical therapy, more income alone may not necessarily help. Further, institutions and technology can change at different times. The fact that more medical advances arrived in the era after World War II does not mean income levels were the cause. Likewise, the fact that people may have allocated healthcare budgets to the new technologies does not mean they did it so because of higher incomes then. They could have spent because the new technologies on offer were more attractive in terms of higher success rates; heart surgery became routinely successful, say, or new cardiovascular treatments produced better results than before, and allocating resources became worthwhile (e.g. Newhouse 1993; Cutler and Kadiyala 2003; Weisbrod 1991). This is not just about the marginal cost of medical treatment, but also about the expected or the perceived private and social benefit from it (Pauly 1986).

Further, the long-term path of non-communicable diseases over the past 150 years undermines the notion that advanced countries such as the UK now spend more on healthcare because they have higher rates of such diseases along with higher incomes. Going by the long-term path of death rates from most such diseases, the rates have been historically very low. They used to be much higher in the past, just that effective medical treatments (to spend on) were unavailable then and have only recently developed, and have become more widely accessible because of institutional change. Thus technological capability—or the elasticity of the supply of effective medical treatments—may be more relevant than income elasticity of demand for why people devote more resources to the matter today than in the past.

The last of the four issues to be fit in is about the likely political roots of healthcare spending. Expressing the desire to age, aging defined as shifting profiles, people may vote their preferences, for instance, about the NHS budget, just like spending on infrastructure grew in the decades after the Second Reform Act and the Public Health reforms in the late nineteenth century (Millward and Sheard 1995; Bell and Millward 1998; Millward and Bell 1998; Millward 2000; Harris 2004, p. 405; Floud et al. 2011, p. 174). For example, back then, nominal loans central government gave to local sanitary authorities—budget allocations— grew four to five times over 1871–1890, and then climbed another sixfold over the next 11 years.

Figure 9.5 shows the share of public health spending (including capital works) in the total spending by the local authorities. This series is made up of outlays on hospitals, sewerage, refuse disposal and water supply. The four were about 6–7% in the early 1880s and grew to about 16% in early nineteenth century. Meanwhile, the share of local government in the total government spending grew from 12.5% in 1820 to 32% in 1870–1871, to 51.1% in 1905 (Szreter 2005, p. 347).

This issue of political roots connects directly with theories that explore factors that boost public spending (Buchanan 1967; Meltzer and Richard 1981). In these frameworks, the backdrop is expansion of the franchise, which is a change of regime from autocracy to one of majority voting. The models then explore the likelier outcome for public spending as the regime changes. The expanding franchise adds voters mostly from the lower end of the income distribution. Being poorer, the new median voter, as portrayed in Downs (1957, 1961), finds redistribution to be more advantageous and thus votes for more public spending.

However, for public healthcare spending at the time the franchise had expanded, the emphasis on the central government as the main driver of spending in these theories may have been incorrect. Although the pressure for better services came

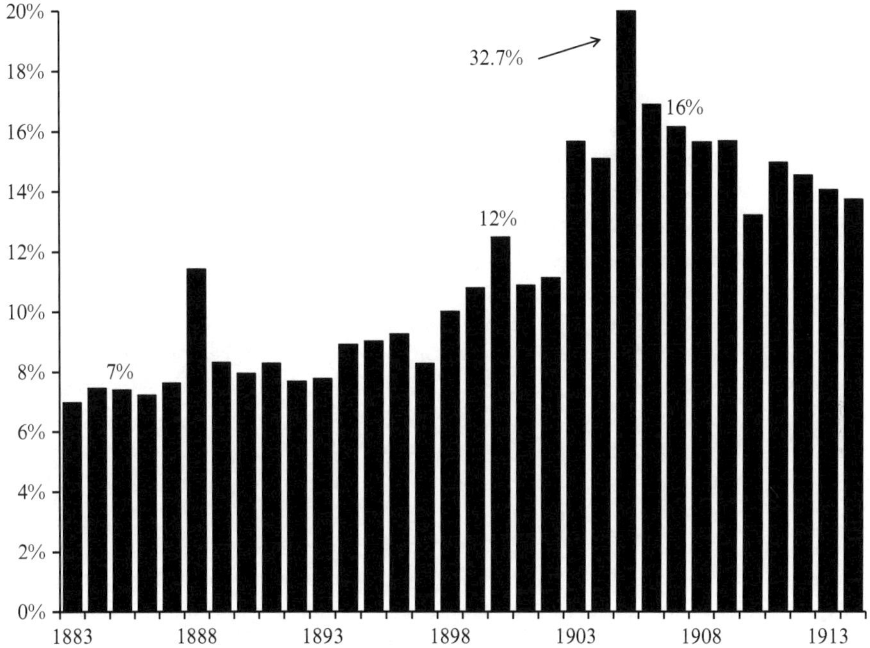

Fig. 9.5 Public health spending as % of total spending by local authorities, including from loans and on capital works, England and Wales, 1883–1914 (Source: Mitchell (2011), pp. 612–619. Note: The large % for 1905 comes from the formation of the Metropolitan Water Board)

from technocrats, citizens, and the central government, most of the increase in productive public spending on roads, sewers, education, police etc. was done by local governments. Finance was not forthcoming from the center. Local property taxes were the main source, later supplemented by local authorities' trading profits. A careful accounting of this issue in Millward and Sheard (1995, p. 501) infers " ..if the nineteenth century was the era of local government, it was also a period when local authorities had to fend for themselves. Central government might establish standards for poor relief, education, policing and public health in the 1860s and 1870s, but it was not willing to use general income taxes or indirect taxes to support or equalize the burden on local communities." Though much of the long-term spending occurred at the local level, it likely varied by the income distribution of the local voters, and thus by the income of the median voter in each locality.

Furthermore, as Chap. 6 pointed out, the type of public health spending back then may not have been of the type where resources raised from one segment of the electorate were transferred or 'redistributed' to others for one-shot consumption. It was more in the form of infrastructure investments that likely had generational ramifications by contributing to the decline of infectious diseases. If the polity is an underlying reason for this type of healthcare spending over time, then it is unclear how some deep-rooted micro-level income elasticity of demand may have played the more pivotal role.

Likewise, for the more recent time, long after 100% franchise was attained, the project Abel-Smith and Titmuss (1956) undertook may have been a technocratic exercise to inform policymakers of the allocations required in the years ahead. This is not an issue of whether the projections turned out right or wrong. It is of how the allocation to the NHS may have been decided. By surmising the prospective costs of the NHS, Abel-Smith and Titmuss may have implicitly prescribed some 'income elasticity' for a given path of per capita incomes, especially if their recommendation had been adopted. This would, however, not be the μ_y construed in the usual sense of individual micro-level demand.

Later on, the budget may have had to adjust to medical advances and to the new aging trends. Abel-Smith and Titmuss (1956), for instance, had no way of knowing the new technologies that would become available over the next two decades when spending grew faster (Fig. 9.3). Neither did they know that children from the transition cohorts, the likely beneficiaries of past spending on healthcare infrastructure, would sway the aging trends. It is not necessary that current income growth since the 1960s caused the technology, the aging, and the budget allocations.

In sum, if four things, (1) the high value an electorate places on healthcare spending especially on basic infrastructure and on universal insurance; (2) policy and budget allocations; (3) the trends in aging; and (4) the medical technology costs; sway allocation to healthcare, then some a priori aggregate-level μ_y may not be very useful for grasping the path of spending. Micro-level ones, for reasons mentioned earlier, may not be suitable either. Would not, then, the assertion in Hall and Jones (2007) "our model makes the prediction that if one looks hard enough and carefully enough, one ought to be able to see income effects [μ_y more than 1] in the micro data" be beside the point?

If one maintains, instead, that conventional μ_y, even the micro-level one, must be relevant, then it is unclear why it would stay invariant over decades. To conjecture, it could be influenced by at least four long-term issues: (1) change of regime (smaller μ_y under universal health insurance versus without); (2) the effectiveness and the relative price of the medical technology on offer (smaller μ_y if, say, stent is routinely successful and relatively cheaper like a safe good you do not forego); (3) income distribution (likely smaller μ_y as income gaps narrow and more people spend on healthcare as something they do not forego); (4) education (smaller μ_y as more people know just when to seek medical attention, and do not take chances by foregoing it; or track their health by regular medical checkups they do not miss). Insurance regime, relative price and safety of effective technology, income distribution, education: based on them if one may guess the path of μ_y over the twentieth century, it may have become progressively more inelastic or less than 1. If so, the notion that a steadfast $\mu_y > 1$ is the reason why the sector's share grew could be off the mark.

None of this, however, makes the choices in Tables 9.1 and 9.2 easier. The first row is no likelier than others. What might be true is that it may be unsafe to label μ_y as income elasticity in the usual sense. Further, if "true" μ_y is really unknowable for current policy or for framing an outlook, then policy must know precisely what makes up the residual so that only the elements that are uncorrelated to aging, or ones that can only boost aging, are the targets of cost containment.

The main deduction in the previous section and this one is that if aging is a key factor in healthcare spending, then aggregate-level policy prescriptions can be socially sub-optimal if they are at odds with the value people place on the matter. Further, the sector's share in the economy may have grown for reasons other than real per capita income growth.

References

Abel-Smith, B. (1967). *An international study of health expenditure* (Public Health Paper No. 32). Geneva: WHO.

Abel-Smith, B., & Titmuss, R. M. (1956). *The cost of the national health service in England and Wales*. London: Cambridge University Press.

Acemoglu, D., & Robinson, J. (2000). Why did the West extend the franchise? Democracy, inequality and growth in historical perspective. *The Quarterly Journal of Economics, 115*(4), 1167–1199.

Acemoglu, D., Finkelstein, A., & Notowidigdo, M. J. (2013). Income and health spending: Evidence from oil price shocks. *Review of Economics and Statistics, 95*(4), 1079–1095.

Akerlof, G. (1970). The market for 'lemons': Qualitative uncertainty and the market mechanism. *The Quarterly Journal of Economics, 84*(3), 488–500.

Anderson, O. W., & Benham, L. (1970). Factors affecting the relationship between family income and medical care consumption. In H. E. Klarman (Ed.), *Empirical studies in health economics*. Baltimore: Johns Hopkins University Press.

American Medical Association. (1978). *The demand for healthcare, report of the national commission on the cost of medical care 1976–77* (pp. 47–93). Chicago: American Medical Association.

Arrow, K. (1963). Uncertainty and welfare economics of medical care. *American Economic Review, 53*, 941–973.

Baumol, W. J. (1988, Fall). Containing medical costs: Why price controls won't work. *The Public Interest, 93*, 37–53.

Baumol, W. J. (1993). Health care, education and the cost of disease: A looming crisis for public choice. *Public Choice, 77*, 17–28.

Bell, F., & Millward, R. (1998). Public health expenditures and mortality in England and Wales 1870–1914. *Continuity and Change, 13*, 285–293.

Blomqvist, A. G., & Carter, R. A. L. (1997). Is health care really a luxury? *Journal of Health Economics, 16*(2), 207–230.

Breyer, F., & Felder, S. (2006). Life expectancy and health care expenditures: A new calculation for Germany using the costs of dying. *Health Policy, 75*(2), 178–186.

Breyer, F., Felder, S., & Costa-Font, J. (2011). *Does ageing really affect health expenditures? If so, why?* www.voxeu.org. 14 May 2011.

Buchanan, J. (1967). *Public finance in democratic processes*. Chapel Hill: University of North Caroline Press.

Culyer, A. J. (1989). Cost containment in Europe. In *Healthcare systems in transition*. Paris: OECD.

Cutler, D. L. (2014). *Quality cure*. Berkeley and Los Angeles, California: University of California Press.

Cutler, D. L., & Kadiyala, S. (2003). The returns to biomedical research: Treatment and behavioral effects. In K. M. Murphy & R. H. Topel (Eds.), *Measuring the gains from medical research: An economic approach*. Chicago: University of Chicago Press.

Cutler, D. L., & Sheiner, L. (1998). *Demographics and medical care spending: Standard and non-standard effects* (NBER Working Paper no. W6866). Cambridge, MA.

Di Matteo, L., & Di Matteo, R. (1998). Evidence on the determinants of Canadian provincial government health expenditures: 1965–1991. *Journal of Health Economics, 17*(2), 211–228.

Dormont, B., Grignon, & Huber, H. (2006). Health expenditure growth: Reassessing the threat of ageing. *Health Economics, 15*(9), 947–963.

Downs, A. (1957). *An economic theory of democracy*. New York: Harper and Row.

Downs, A. (1961). Problem of majority voting: In defense of majority voting. *Journal of Political Economy, 69*(2), 192–199.

European Commission. (2012a). *The 2012 ageing report: Underlying assumptions and projection methodologies.*

European Commission. (2012b). *The 2012 ageing report: Economic and budgetary projections for 27 EU States (2010–2060).*

Felder, S. M., & Schmidt, H. (2000). Health care expenditure in the last months of life. *Journal of Health Economics, 19*, 679–695.

Feldstein, M. (1971). Hospital cost inflation: A study of non-profit price dynamics. *American Economic Review, 61*, 853–872.

Floud, R., Fogel, R. W., Harris, B., & Hong, S. C. (2011). *The changing body, health nutrition and human development in the western world since 1700* (NBER). New York: Cambridge University Press.

Fogel, R. W. (1999). Catching up with the economy. *American Economic Review, 80*, 1–21.

Fuchs, V. R. (1990). The health sector's share of the gross national product. *Science, 247*, 534–538.

Fuchs, V., & Kramer, M. (1972). Determinants of expenditures for physicians' services in the US 1948–1968. In V. Fuchs (Ed.), *The health economy*. Cambridge: Harvard University Press, 1982.

Gerdtham, U., Sogaard, J., Anderson, F., & Jonsson, B. (1992). An econometric analysis of health care expenditure: A cross-section study of the OECD countries. *Journal of Health Economics, 11*(1), 63–84.

Getzen, T., & Poullier, J. P. (1992). International health spending forecasts: Concepts and evaluation. *Social Science and Medicine, 34*(9), 1057–1068.

Getzen, T. (2000). Health care is an individual necessity and a national luxury: Applying multi-level decision models to national healthcare expenditure. *Journal of Health Economics, 19*, 259–270.

Hall, R., & Jones, C. (2007). The value of life and rise in health spending. *The Quarterly Journal of Economics, 122*, 39–72.

Harris, B. (2004). *The origins of the British welfare state: Social welfare in England and Wales 1800–1945*. New York: Palgrave, Macmillan.

Hogan, C., Lunney, J., Gabel, J., et al. (2001). Medicare beneficiaries' cost of care in the last year of life. *Health Affairs (Milwood), 20*(4), 188–195.

Kleiman, E. (1974). The determinants of national outlay on health. In M. Perlman (Ed.), *The economics of health and medical care*. London: Macmillan.

Levit, K. R. (1982). Personal health expenditures by state, selected years 1966–1978. *Health Care Financing Review, 4*(2), 1–16.

Lubitz, J., & Prihoda, R. (1984). The use and cost of medicare services in the last two years of life. *Healthcare Financing Review, 5*(3), 117–131.

Lubitz, J., & Riley, G. F. (1993). Trends in medicare payments in the last year of life. *The New England Journal of Medicine, 328*(15), 1092–1096.

Manning, W., et al. (1987). Health insurance and the demand for medical care: Evidence from a randomized experiment. *American Economic Review, 77*, 252–277.

Maxwell, R. (1981). *Health and wealth: An international study of health care spending*. Lexington: Lexington Books.
Meltzer, A. H., & Richard, S. F. (1981). A rational theory of the size of government. *Journal of Political Economy, 89*(5), 111–118.
Miller, T. (2001). Increasing longevity and medicare expenditures. *Demography, 38*(2), 215–226.
Millward, R. (2000). Urban government, finance, and public health in Victorian Britain. In R. Morris & R. Trainor (Eds.), *Urban governance: Britain and beyond since 1750*. Aldershot: Ashgate.
Millward, R., & Bell, F. (1998). Economic factors in the decline of mortality in late 19th and 20th century Britain. *European Review of Economic History, 2*, 263–288.
Millward, R., & Sheard, S. (1995). The urban fiscal problem, 1870–1914: Government expenditure and finance in England and Wales. *Economic History Review, 2*, 501–535.
Murphy, K. M., & Topel, R. H. (2003). The economic value of medical research. In K. M. Murphy & R. H. Topel (Eds.), *Measuring the gains from medical research: An economic approach*. Chicago: University of Chicago Press.
Newhouse, J. P. (1977). Medical care expenditure: A cross national survey. *Journal of Human Resources, 12*(1), 115–125.
Newhouse, J. P. (1987). Cross-national differences in health spending: What do they mean? *Journal of Health Economics, 6*, 159–162.
Newhouse, J. P. (1992). Medical care cost: How much welfare lost? *Journal of Economic Perspectives, 6*(3), 3–21.
Newhouse, J. P. (1993). An iconoclastic view of health cost containment. *Health Affairs, 12*(suppl. 1), 152–171.
Newhouse, J. P., & Phelps, C. E. (1976). New Estimates of price and income elasticities of medical care services. In R.N. Rosett (Ed.), *The Role of Health Insurance in the Health Services Sector*. NBER: New York.
Nordhaus, W. D. (2005). Irving Fisher and the health of nations. *American Journal of Sociology and Economics, 64*(1), 367–392.
OECD. (2013). *Public spending on health and long-term care: A new set of projections* (OECD economic policy papers, no. 6). Paris: OECD.
Okunade, A. A., & Murthy, V. N. R. (2002). Technology as a major driver of healthcare costs: A cointegration analysis of the Newhouse conjecture. *Journal of Health Economics, 21*, 147–159.
Oliveira Martins, J., & de la Maisonneuve C. (2006). The drivers of public health expenditure on health and long-term care: An integrated approach. *OECD Economic Studies* No. 43, 2006/2.
O'Neill, C., Groom, L., Avery, A. J., et al. (2000). Age and proximity to death as predictors of GP care costs: Results from a study of nursing home patients. *Health Economics Letters, 4*, 3–9.
Parker, S., & Wong, R. (1997). Household income and healthcare expenditures in Mexico. *Health Policy, 40*, 237–255.
Pauly, M. V. (1986). Taxation, health insurance, and market failure in the medical economy. *Journal of Economic Literature, 24*(2), 629–675.
Reinhardt, U. E. (2003). Does the aging of the population really drive the demand for health care? *Health Affairs, 22*(6), 27–39.
Riley, G. F., Lubitz, J., Prihoda, R., et al. (1987). The use and costs of medicare services by cause of death. *Inquiry, 24*, 233–244.
Scheiber, G. (1990). Health expenditures in major industrialized countries, 1960–1987. *Health Care Financing Review, 11*(4), 159–167.
Schwartz, W. B. (1987). The inevitable failure of cost containment strategies: Why they can provide only temporary relief. *Journal of American Medical Association, 257*, January: 220–224.
Seshamani, M., & Gray, A. (2002). The impact of ageing on expenditures in the National Health Service. *Age and Ageing, 31*, 287–294.

Seshamani, M., & Gray, A. (2004). A longitudinal study of the effects of age and time to death on hospital costs. *Journal of Health Economics, 23*, 217–235.

Szreter, S. (2005). *Health and wealth: Studies in history and policy*. Rochester: University of Rochester Press.

Wagstaff, A., van Doorslaer, E., & Paci, P. (1991). Equity in finance and delivery of health care: Some tentative cross-country comparisons. In A. McGuire, P. Fenn, & K. Mayhew (Eds.), *Providing health care*. Oxford: Oxford University Press.

Weisbrod, B. A. (1991, June). The healthcare quadrilemma: An essay on technological change, insurance, quality of care, and cost containment. *Journal of Economic Literature, 29*, 523–552.

Werblow, A., Felder, S., & Zweifel, P. (2007). Population ageing and health care expenditure: A school of 'red herrings'? *Health Economics, 16*, 1109–1126.

Zweifel, P., Felder, S., & Meiers, M. (1999). Ageing of population and health care expenditure: A red herring? *Health Economics, 8*, 485–496.

Chapter 10
The Transitions of Aging

10.1 Seen as Profiles of Illness and Deaths

The $\beta(t)$s studied so far have been for the aging profiles. Such profiles likely couple with illness profiles that have their own $\beta(t)$s. That aging may have swayed spending does not necessarily mean illness did so in the same way and by as much. Their influence would be identical only if the illness and the aging profiles overlap perfectly and their $\beta(t)$s correlate 1. Their overlap cannot be perfect because deaths are usually a subset of illness, and the levels of the profiles should differ. The perfect correlation cannot be supposed as all illnesses do not end in deaths, and their trends can vary.

Focusing on adult-age non-communicable diseases at the aggregate-level over the long term, this chapter examines analytically period-profiles of spending, of illness and of aging in six main scenarios: (1) expansion of deaths and illness; (2) compression of deaths, but expansion of illness; (3) compression of deaths as well as of illness; and (4) the influence of norms, technology, and prevention; (5) inter-temporal links in profiles; (6) a combination of norms, technology and inter-temporal links.

Segments of some of the scenarios were stated in Chap. 9. They are crystallized here in terms of profile parameters $\alpha(t)$ and $\beta(t)$ so that assumptions are clearer and they can be related to the historical debate on the trends in illness and death rates. My goal is to explore healthcare spending profiles, and their path the scenarios could be capable of generating. The scenarios are stripped bare, however. They only have profile parameters, and costs that embody technology. There is no utility function, discount rate, risk aversion parameter, uncertainty, information asymmetries, and budget constraints; just the basic total costs and marginal costs, the shifting profiles, and their ramifications for spending. I will, however, indicate in some places where issues such as markups above marginal costs can appear when medical technology is supplied under imperfect competition.

S. Arora, *The Transitions of Aging*, International Perspectives on Aging 12,
DOI 10.1007/978-3-319-14403-0_10

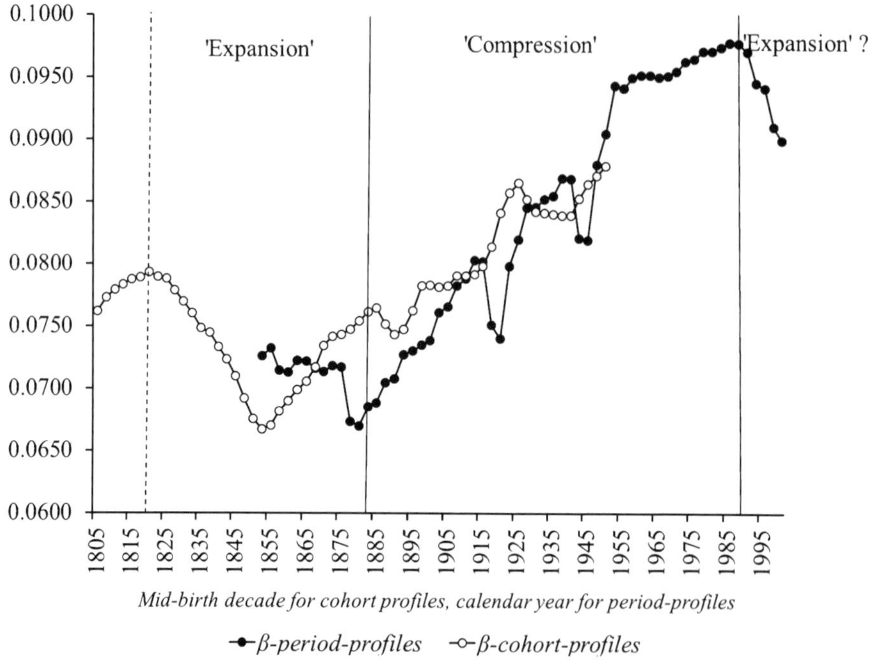

Fig. 10.1 $\beta(t)$ of Cohort- and period-profiles, 1805–2000

While exploring the scenarios, it helps to keep a longer term view like Fig. 10.1 that shows $\beta(t)$s of cohort- and period-profiles of aging since the nineteenth century. Given the format of the study, it may be clear by now that five cohorts contribute to form the $\beta^C(t)$ segment of each period-$\beta(t)$. Except for brief intervals of the two world wars, Fig. 10.1 reveals three main long-term eras. The first one labeled expansion is roughly the first three quarters of the nineteenth century. The compression runs since about the end of the nineteenth century or early-twentieth century through the mid-1980s. The pattern since then seems like expansion again though it is unclear if it is an early stage of a new longer-term trend or something that may reverse like the brief dips during the world wars. The main questions in Fig. 10.1 are: what turns the trends to form the transition of aging, and what would be the healthcare spending as the transitions happen.

To develop the scenarios, some variables need to be defined. For each year t and adult age a, there are a series of period-profiles of aging $d(t, a)$ defined by death rates, and of illness $i(t, a)$ defined by illness rates. Death *rates* across age groups in each profile sum up to the total death rate. Similarly, age-specific illness rates in the illness profiles add up to the total illness rate. Both profiles thus pertain to per capita measures. Subsumed in $i(t, a)$ is duration of illness, which can vary for reasons other than incidence such as intensity of illness, time required for medical treatment, time prescribed for recovery, sick leave policies, among others.

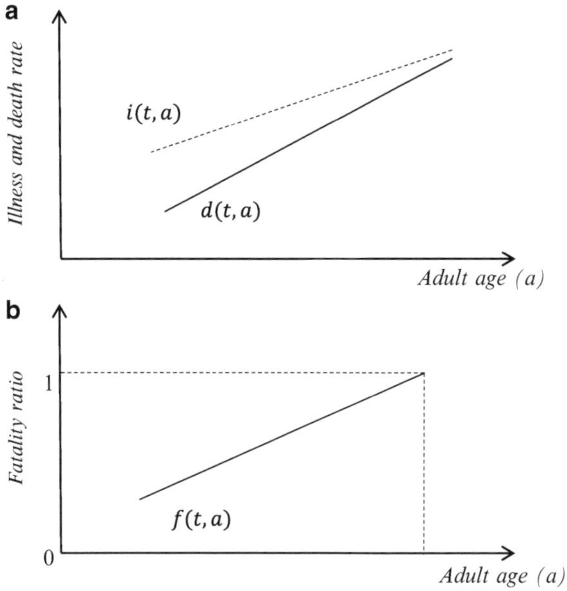

Fig. 10.2 Stylized age-profiles (**a**) Illness and death rates (**b**) Fatality ratios

Both types of profiles slope upward as illness rates and death rates climb with age. The parameters of aging profiles $d(t, a)$ are $\alpha_d(t)$, $\beta_d(t)$. The parameters of illness profiles $i(t, a)$ are $\alpha_i(t)$, $\beta_i(t)$. It should be clear that $\beta_d(t)$, the slope of $d(t, a)$, is the first derivative of $d(t, a)$ with respect to age; likewise, $\beta_i(t)$ is the slope of $i(t, a)$. To keep things general, I will be using slopes (β) symbolically without assuming a specific functional form for the profiles.

Assume $\alpha_d(t), \beta_d(t), \alpha_i(t), \beta_i(t) \in (0, \infty)$; i.e. they are all positive, but less than infinite. It means the profiles $d(t, a)$ and $i(t, a)$ are neither horizontal nor vertical, ruling out two theoretical outcomes: (1) vertical $d(t, a)$, $i(t, a)$, the ideal case in which nobody ever falls ill or dies before reaching the limit to the human lifespan; (2) horizontal $d(t, a)$, $i(t, a)$, where death and illness rates are constant irrespective of age. Both profiles always slope upward, steeper in some periods than in others.

Assume $d(t, a) \leq i(t, a)$ since $d(t, a) \subseteq i(t, a)\ \forall\ t, a$: death rates are always less than or equal to illness rates since deaths come from a given pool of illness episodes. Further, with finite lifespan, the series of $\alpha_d(t)$ and $\beta_d(t)$ relate inversely; likewise for $\alpha_i(t)$ and $\beta_i(t)$. With death rates less than or equal to illness rates, and with the inverse relation of $\alpha(t)$ and $\beta(t)$, it turns out that $\alpha_d(t) \ll \alpha_i(t)$, and $\beta_d(t) \gg \beta_i(t)$. Otherwise, one can stumble into an awkward situation where the death rate is more than the illness rate. The consequence is that the two profiles are not parallel, like the ones shown in Fig. 10.2a.

Further, downward shifts in $i(t, a)$, or a fall in illness rates, coincide with steeper slopes or growing $\beta_i(t)$; upward shifts, or more illness rates, mean a flatter slope and falling $\beta_i(t)$. That is, as illness rates grow, $i(t, a)$ shifts upward, $\Delta i(t, a) > 0$, and $\Delta \beta i(t) < 0$ as the new slope is flatter than the one before.

Moreover, define an age-profile of fatality ratios, death rate per illness rate, or deaths per illness, as:

$$f(t,a) = \frac{d(t,a)}{i(t,a)}, \tag{10.1}$$

so that $f(t, a) \in (0, 1]$, for $d(t, a) > 0$ and $i(t, a) > 0, \forall\, t, a$. As long as death rates and illness rates are positive, and death rates are less than or equal to illness rates, the fatality ratio is always more than 0, but never more than 1 as deaths from non-communicable diseases in aggregate do not exceed the illness they cause. This framework is incapable of studying cases where some disease causes illness $i(t, a) > 0, \forall\, t, a$, but no deaths result from it. Keeping $d(t, a) > 0$ is an appropriate assumption for the aggregate level, though not necessarily for specific diseases.

Given the assumptions about $d(t, a)$ and $i(t, a), f(t, a)$ too slopes upward with age, but stays within (0, 1]. Illness is increasingly difficult to treat with age, or illness is more fatal as people age, and the fatality ratio is greater for older people; hence, Fig. 10.2b. A fall in the ratios is a downward shift that leaves $f(t, a)$ steeper than before as it stays tethered near 1 at extreme old age.

Further, a key implicit assumption of this framework is that people know when they are ill and the type of illness they have, not something to take lightly (Arrow 1963; Akerlof 1970; Sen 1998). The discussion below will point out when the issue can be critical. One of the scenarios has a diagnostic technology that lets people know what type of illness they have, which prompts them to seek medical remedy or treatment. The diagnostics and treatments are associated with costs. The scenario then develops healthcare spending profiles as functions of illness rates, aging, diagnostic costs, treatment costs, among others, shedding some light on the results in Chap. 9, and on the issue of cost containment.

The scenarios have four building blocks. The first one is the expansion of deaths, when initial states of $d(t, a)$s grow and their slopes become flatter:

$$\alpha_d(t+i) < \alpha_d(t+i+1) \; and \; \beta_d(t+i) > \beta_d(t+i+1), \qquad i = 0 \ldots T \tag{10.2a}$$

The second is compression of deaths, when initial states fell and the profiles $d(t, a)$ became steeper over the long run:

$$\alpha_d(t+i) > \alpha_d(t+i+1) \; and \; \beta_d(t+i) < \beta_d(t+i+1), \qquad i = 0 \ldots T \tag{10.2b}$$

To my knowledge, there is no annual aggregate-level record for illness rates like the one for the aging profiles. The chapter can examine illness profiles only analytically. It will, however, discuss some findings on select sub-populations scholars have explored, as well as some recent trends of illness in national surveys. The expansion of illness rates can be written as:

$$\alpha_i(t+i) < \alpha_i(t+i+1) \; and \; \beta_i(t+i) > \beta_i(t+i+1), \qquad i = 0 \ldots T \tag{10.3a}$$

and their compression as:

$$\alpha_i(t+i) > \alpha_i(t+i+1) \; and \; \beta_i(t+i) < \beta_i(t+i+1), \qquad i = 0 \ldots T. \quad (10.3b)$$

These four building blocks can now be combined to develop various scenarios.

Expansion of deaths and of illness rates: Consider the case for the segment of the nineteenth century labeled expansion in Fig. 10.1. It pertains to Eq. 10.2a. But which scenario of the illness rates went with it, expansion (Eq. 10.3a), or compression (Eq. 10.3b)?

For $\beta_d(t)$ to have shrunk for several decades, there must have been a continual flow of illnesses. That is, $\beta_i(t)$ may have shrunk as well. Otherwise, sooner or later, death rates in the profiles may have stopped growing, and $\beta_d(t)$ would reverse direction. For instance, if death rates from heart disease were growing, then illness rates from heart disease must have been climbing as well. A period of inflection in $\beta_d(t)$—a turning point, as one did arrive—would reflect that the flow of illnesses may have slowed.

In this scenario the issue of whether people really know they are ill is vital. When death rates are rising, the false self-perception of not being ill may result in a misleading trend of the reported illness rates (Sen 1998). Under-reported illness would thus be a serious concern. The reliable trend in illness rates would be one that at least reaffirms the growing death rate, if not presages its growth.

It is likelier, then, for this period to have been one of expansion of illness and of deaths: a series of shrinking $\beta_d(t)$ and $\beta_i(t)$, or a series of flatter profiles of both—an unlikely basis for healthy aging. The combination can be written as, expansion of deaths:

$$\alpha_d(t+i) < \alpha_d(t+i+1) \; and \; \beta_d(t+i) > \beta_d(t+i+1), \qquad i = 0 \ldots T \quad (10.4a)$$

and expansion of illness

$$\alpha_i(t+i) < \alpha_i(t+i+1) \; and \; \beta_i(t+i) > \beta_i(t+i+1). \qquad i = 0 \ldots T. \quad (10.4b)$$

To see what might happen to the trend of the fatality ratio, take natural logs on both sides of Eq. 10.1 and differentiate with respect to time (suppressing a):

$$g_f(t) = g_d(t) - g_i(t), \quad (10.5a)$$

where, g_f, g_d, and g_i are percentage changes in the aggregated fatality ratio, death rates, and illness rates. There are three possibilities for $g_f(t)$:

$$g_f(t) \gtreqless \begin{cases} g_d(t) > g_i(t) \\ g_d(t) = g_i(t) \\ g_d(t) < g_i(t) \end{cases} \quad (10.5b)$$

The scenarios below will repeatedly refer to Eq. 10.5b. For the current one, all else equal, the top term on right-hand-side of Eq. 10.5b does not seem viable over the very long run: all else equal, one may run out of illness to feed the faster-growing deaths (both are positive in expansion). The middle and the third ones are likelier. That is,

$$\frac{d(t+i,a)}{i(t+i,a)} = \frac{d(t+i+1,a)}{i(t+i+1,a)}, or, \frac{d(t+i,a)}{i(t+i,a)} > \frac{d(t+i+1,a)}{i(t+i+1,a)}, \quad (10.5c)$$

$i = 0 \ldots T, a = j \ldots N$. As a result, the ratio could either be trendless (the illness and death rates grow at about the same average rate over the longer run), or it could trend down (illness rates grow faster than death rates). 'Trendless' should not be seen as a constant, but as fluctuation around some constant with no persistent deviation from it. So, the fatality ratio's profile (Fig. 10.2b) could have either stayed static, or a series of fatality profiles had shifted down and become steeper. Static or unchanged profile would be $(f(t+i,a) = f(t+i+1,a))$. A profile that has shifted down would be $(f(t+i,a) > f(t+i+1,a))i = 0 \ldots T, a = j \ldots N$; and since they slope up with age, in each shifted profile

$$f(t+i,j) - f(t+i+1,j) > \ldots > f(t+i,N) - f(t+i+1,N), \quad (10.5d)$$

or, in absolute terms, the ratios falls at each age, but more so for the youth (e.g. the first term) than for older people (e.g. the last term) for whom the ratio is closer to 1 to begin with. Thus, steeper profile.

Shrinking fatality ratios are usually seen as a good outcome since they indicate fewer deaths per illness. However, this scenario cautions that the underlying trend of illness and deaths should be examined. They could both be growing. If so, the falling ratio would be a good thing, but in a bad situation.

That horrible episode 'gave way' to the era of compression of deaths (the 'transitions' explored later). They could have been accompanied by three possible trends in illness profiles: static (unchanged), shifted down (compression), shifted upward (expansion).

The first combination of the *compression of deaths, but static illness rates* can be written as:

$$\alpha_d(t+i) > \alpha_d(t+i+1) \; and \; \beta_d(t+i) < \beta_d(t+i+1), \qquad i = 0 \ldots T, \quad (10.6a)$$

$$\alpha_i(t+i) = \alpha_i(t+i+1) \; and \; \beta_i(t+i) = \beta_i(t+i+1). \qquad i = 0 \ldots T. \quad (10.6b)$$

Since only death rates fall,

$$\frac{d(t+i,a)}{i(t+i,a)} > \frac{d(t+i+1,a)}{i(t+i+1,a)} i = 1 \ldots T, a = j \ldots N, \quad (10.6c)$$

or, the fatality ratio drops: $f(t+i,a) > f(t+i+1,a)$. Moreover, since each profile slopes upward with age,

$$f(t+i,j) - f(t+i+1,j) > \ldots > f(t+i,N) - f(t+i+1,N), \quad (10.6d)$$

which is like the outcome in Eqs. 10.5c and 10.5d, but with strict inequality. Here, too, the falling ratio can be misleading as illness rates have not budged; only deaths have shrunk. Since $g_i(t) = 0$ and death rates are falling, the middle-term and the first term on the right-hand-side of Eq. 10.5b are irrelevant; the last term shapes the falling trend.

The next combination is *compression of deaths* and *expansion of illness*:

$$\alpha_d(t+i) > \alpha_d(t+i+1)\,and\,\beta_d(t+i) < \beta_d(t+i+1), \qquad i = 0\ldots T \quad (10.7a)$$

$$\alpha_i(t+i) < \alpha_i(t+i+1)\,and\ \beta_i(t+i) > \beta_i(t+i+1). \qquad i = 0\ldots T. \quad (10.7b)$$

The outcome for the fatality ratios is similar to the one in the previous case, but the ratio would drop more in this case as illness rates (the denominators) grow; so the ratio's profiles shift down more and become steeper than in the previous case. In both scenarios, falling deaths may convey improving health in a limited sense: people live longer, but not with less illness than before.

The last pairing of *compression of deaths* is with the *compression of illness*. The initial states of both illness and death-rate profiles shrink and their slopes become progressively steeper. This case is what scholars tend to have in mind when they use period-life-expectancy or falling long-term death rates to convey or imply that health must have improved too. The usual pretext is that the long-term falling trend of deaths must correlate strongly positive with the trend of illness rates. It can be written as:

$$\alpha_d(t+i) > \alpha_d(t+i+1)\,and\ \beta_d(t+i) < \beta_d(t+i+1), \qquad i = 0\ldots T \quad (10.8a)$$

$$\alpha_i(t+i) > \alpha_i(t+i+1)\,and\ \beta_i(t+i) < \beta_i(t+i+1), \qquad i = 0\ldots T \quad (10.8b)$$

To grasp the trend of fatality ratios, keep in mind that both g_d and g_i in Eq. 10.5b are now negative (pick numbers such as -5% or -3% to make it easier). Death rates may fall faster than illness rates for a long time, for instance, if medical technology can continually prevent deaths, though usually less for very old adults. The ratio's profile shifts down and become steeper than before; or,

$$\frac{d(t+i,a)}{i(t+i,a)} > \frac{d(t+i+1,a)}{i(t+i+1,a)}, \qquad (10.8c)$$

$i = 0\ldots T, a = j\ldots N$. The profile of the fatality ratios shift down $f(t+i,a) > f(t+i,a)$ $i = 0\ldots T, a = j\ldots N$. And, like in other scenarios, since each the profile slopes upward with age,

$$f(t+i,j) - f(t+i+1,j) > \ldots > f(t+i,N) - f(t+i+1,N), \qquad (10.8d)$$

which is similar to Eq. 10.6d, except this one pertains to falling illness rates instead (comparisons to be discussed in Table 10.1 shortly).

For the term in the middle of the right-hand-side of Eq. 10.5b the falling long-term $g_d(t)$ and $g_i(t)$ are identical (e.g. -4% rate each), and the ratio is trendless or $g_f(t) = 0$. This can be expressed as

Table 10.1 Scenarios of death and illness rate period-profiles

	Scenario	Trend $f(t, a)$	Good/Bad?
I	Expansion of death and illness rates		
	(i) $g_d(t) = g_i(t)$	Trendless, $g_f(t) = 0$	Bad
	(ii) $g_d(t) < g_i(t)$, both growing	Falling, $g_f(t) < 0$	Bad
II	Compression of death rates, but static illness rates	Falling, $g_f(t) < 0$	Bad
III	Compression of death rates, but expansion of illness rates	Falling, $g_f(t) < 0$, but more than II	Bad
IV	Compression of death rates and illness rates		
	(i) $g_d(t) = g_i(t)$	Trendless, $g_f(t) = 0$	Good
	(ii) $g_d(t) > g_i(t)$, both shrinking	Falling, $g_f(t) < 0$	Best

$$\frac{d(t+i,a)}{i(t+i,a)} = \frac{d(t+i+1,a)}{i(t+i+1,a)}, \tag{10.8e}$$

$i = 0 \ldots T, a = j \ldots N$. Meanwhile, the ratio's age-profile stays static or $f(t+i,a) = f(t+i,a)$ $i = 0 \ldots T, a = j \ldots N$, but since it grows with age,

$$f(t+i,j) - f(t+i+1,j) > \ldots > f(t+i,N) - f(t+i+1,N), \tag{10.8f}$$

Now think about the last term on the right-hand-side of Eq. 10.5b for this scenario. Can illness rates forever fall faster than death rates over the long term? After a point, if there is nobody falling ill, there might be no deaths either. That does not seem to have been the aggregate-level experience for non-communicable diseases, especially in the context of aging. It would also violate the assumptions: $d(t, a) > 0$, $i(t, a) > 0$ and $d(t, a) \leq i(t, a)$ as $d(t, a) \subseteq i(t, a)$; the profile parameters $\in (0, \infty)$; and that $f(t, a) \in (0, 1]$ has a positive slope. At the micro-level, however, it can stand for some infectious diseases that have been contained, but such a case is outside the scope of an aggregate-level setup for non-communicable diseases.

Table 10.1 lists the basic outcomes for the trend in the fatality ratio under various scenarios. The result that jumps out is that the ratio's trend, by itself, can be very misleading. It can be trendless or be falling in totally opposite situations such as I and IV. Even bad scenarios II and III result in a falling trend, but it falls faster in III than in II. That makes it all the more important, then, to focus on the fundamentals of illness rates and death rates to know the underlying scenario. The ratio, as will become clear shortly, does not have a clear direction when opposing forces are operating. The scenarios, nonetheless, help navigate some critical matters in the historical debate about the trends of illness rates and death rates.

10.2 That Shed Light on a Historical Debate

The issue of their long-term trends in the Transition was raised in an influential article titled "The Failures of Success" by Gruenberg (1977), which argued that in the past severe death rates from infectious diseases, especially in children, may have prevented non-communicable ones from being prominent since few people may have aged enough to suffer from such diseases. Curbing infectious diseases (the success) may have had the unfortunate result of boosting illness and deaths from non-communicable ones (the failure).

The evidence on death rates, however, conveys the opposite. The era when the rates of infectious diseases were severe (failure) was also the era when the rates of non-communicable diseases were very high (failure). Both categories shrank successfully over the Transition, though not together. Further, the era when rates of infections were acute in cohorts of children was also the time those cohorts had very high death rates from non-communicable disease as they aged (Chap. 3). Though this link is an intertemporal one, since the regime had persisted for a long time, it produces high contemporaneous period-rates of both categories in the nineteenth century, which, at least for the death rates, is the reverse of Gruenberg's thesis.

Nonetheless, when the Transition's original framework was proposed in the 1970s, his argument prompted a debate about the relation between the trends in illness and deaths. Perhaps because there was little evidence to the contrary, there was a widespread presumption until then that growing period-life expectancies and falling death rates were accompanied by less illness rates as well (scenario IV in Table 10.1). His thesis, however, implied they could relate inversely, akin to scenario III, possibly to scenario II.

Consequently, scholars began formulating indexes that adjusted life expectancy for disability rates, based on the premise that period-life expectancy stood for lifespan or some measure of years lived. Aside from period-life-expectancies, scholars also began assessing period-health expectancies, as surveys during the 1970s and 1980s often revealed persistent illness rates, recurring functional limitations, and disabilities of various types, even as death rates had shrunk (e.g. Verbrugge 1984; Crimmins et al. 1989, 1994, 1996; Olshansky et al. 1991; Murray and Lopez 1996; Reynolds et al. 1998; Robine et al. 2003).

Unfortunately, annual data on illness at the aggregate level, especially over the very long-term studied in this book, has not been recorded. Scholars have, nonetheless, studied select sub-populations such as health insurance records kept by friendly societies, sickness records kept by trade unions, among others.

As the debate grew and spilled over into the issue of historical trends since the nineteenth century, a series of studies that proved controversial were Riley (1987, 1989, 1991, 1997; Riley and Alter 1989). They examined records of one of the largest English friendly societies, the Ancient Order of Foresters, over 1870–1914, finding that incidence of sickness episodes fell, and sickness durations grew, resulting in more prevalence during a time when overall death rates were falling—the inverse relation of illness and death rates. While it may be improper to compare the finding to an analysis based on period-profiles, what makes the issue compelling is that such evidence is akin to scenario III, but it was occurring at a time when the regime in Fig. 10.1 was likely closer to scenario I, perhaps transitioning to II or IV.

One explanation could be that he may have been referring to overall death rates, and not specifically to the death rates from non-communicable diseases. The population-wide death rates (all disease-categories combined) were shrinking, but underneath the surface, non-communicable disease death rates were climbing (Chaps. 2 and 4). It could be that the growing prevalence may have pertained to non-communicable diseases. In which case, the inverse relation may have been a consequence of comparing apples (prevalence of non-communicable disease) to oranges (overall death rates).

Further, as Sen (1998) has pointed out, reported illness can be unreliable when death rates turns out more than illness rates. Likewise, over time, if deaths are growing, reported-illness rates can be reliable only if they bolster the case for climbing death rates. For example, if increasingly more people were succumbing to heart disease, but reported-illness trends conveyed that heart diseases was not an issue, which one would be closer to the actuality of heart disease, illness rate or death rate? It should be surprising, then, if prevalence rates were not growing along with population-wide death rates from non-communicable diseases. Provided an apples-to-apples comparison were made, what Riley (1989) discovered for the period may not have been a case of an inverse relation, but a case of a positive relation (scenario I). If, however, deaths from such diseases were instead falling in the records of those societies, then it is unclear whether their experience could shed light on what the general population was experiencing.

Further, the main hypotheses for the inverse relation offered were (1) insult accumulation; (2) differential frailty; and (3) beneficial effects of medical attention on the survival from fatal diseases (Alter and Riley 1989). If children who had endured the harsh epidemiologic situation of the nineteenth century were contributing to the rising rates of non-communicable diseases as adults, then an intertemporal link, the long reach of childhood development, may have been a crucial reason as well.

Moreover, profound socio-economic and political changes were occurring at the time, and it may be unsafe to interpret the trends in purely biological terms. The intertemporal links likely had an economic and political dimension as well (Chaps. 6, 7, 8). Further, an aggregate-level case for beneficial effects of medical attention is harder to make for that period of growing death rates. Nonetheless, it can be argued that medical attention may have played a role in two ways: (1) access was unsymmetrical across social classes and medical advice benefited a smaller segment of the population; in which case, though, it may not be a big factor for the aggregate level; (2) there is a counterfactual trend for the death rate which would have been worse without the advice, but that is difficult to establish.

Scholars have explored data on other friendly societies to study other places had experienced similar trends. For example, Edwards et al. (2003) found for the Hampshire Friendly Society that sickness rates overall, not just duration or incidence, had grown over 1868–1921, indicating that illness trends may have varied by region (also see Gorsky et al. 2004, 2011; Harris et al. 2009, 2012). However, the diseases linked to episodes of illness in these studies may only partially match diseases in the aggregate-level data on death rates, for two reasons: (1) not all diseases are fatal; or (2) geographic coverage may be limited so that region-specific issues govern what types of diseases dominate regional trends. Nonetheless, data on select populations have a distinct advantage: they provide a much better idea of trends in non-fatal diseases than what aggregate-level data on death rates alone can provide.

Another vein of research, but at the micro-level, studied medical and anthropometric records of select populations such as Civil War and Vietnam War veterans in the US to grasp the very long-term change in the incidence, sickness durations, prevalence, age of onset, and functional limitations of middle-to-older aged men. They found sizable improvements along all of those dimensions, each related to better anthropometric measures such as stature, body mass index, birth weight, waist-to-hip ratios, among others (e.g. Steckel and Costa 1997; Costa 2002, 2003). For UK, some studies have explored cohort-level or cross-sectional information extending over early-to-mid twentieth century. They too find telling links with anthropometric measures such as stature, though the linkages do not extend to all types of diseases, some showing no relation at all (e.g. Batty et al. 2009).

Critics of this research on time trends, meanwhile, have pointed out that interpreting trends correctly requires paying close attention to institutional, socio-economic, and cultural changes playing out as the data were being recorded. For instance Murray (2003) points out that data on insurance claims are subject to change in the financial health of entities offering insurance, to incentives for claiming illness, and to thresholds of illness used for validating claims. Similarly, data in micro-level studies on very distant cohorts are susceptible to missing major reversals in the middle.

Further, since new institutions for healthcare were developing over this period, it may be unsafe to interpret the data in biological terms such as frailty if institutional changes had triggered some of the trends. On this issue, Johansson (1991) warned very early in the debate that illness data can also be 'inflated' over time. Changing cultural and social norms may govern what is seen as illness. For instance, what was seen as a trivial symptom at time t may come to be viewed as a critical symptom at $t + 1$. The result is that either illness rates would have been under-reported before $t + 1$, as people may have been unaware of the seriousness of the situation, or would be over-reported or reported to reflect the situation more accurately after then. The consequent upward trend is then a result of a change in the norm, perhaps better self-perception of illness, not necessarily some biological or physiological change. This tendency could have been boosted by more frequent diagnoses than before as people had more medical checkups after the National Insurance Scheme widened access to general practitioners, and after the NHS enabled even wider access to medical care. These two themes—availability or access to insurance, and the coverage of insurance, along with new diagnostic and treatment technology—have been prominent in the debate on healthcare spending over the past few decades.

10.3 Norms and Technology Boost Spending

Their ramifications can be studied in the framework of the profiles. But before exploring them, it is better to find out the scenario in Table 10.1 where the issue of inflation, or growing reported illness rates, would be most critical. In scenario I, deaths are growing anyway. It may not matter so much if a new norm inflates illness rates. Inflation may even be welcome if it portrays the situation more accurately than before; deflated illness rates would be the more acute concern here.

The concern about inflation is likelier to be relevant when deaths are compressing. All else equal, death rates fall, but the illness rates convey that deaths should have been growing, or at least not falling by as much. Among the scenarios in Table 10.1, then, the matter seems most relevant to scenario III. In scenario IV, illness rates are shrinking as is; though they may still be inflated, but as long as they indicate the correct long-term direction, the issue may not be as critical, unless inflation reverses the erstwhile falling trend of illness rates—in which case it would become scenario III.

How might inflation play out in scenario III? What would make more people report they have an illness or a potential illness requiring medical attention or treatment? It is best to start with scenario II and then slip into III. Consider only the $\beta(t)$s of the profiles for 3 periods, where the compression of the aging-profiles begins in period 3:

$$\beta_d(1) = \beta_d(2) < \beta_d(3) \tag{10.9a}$$

Likewise, say that $\beta_i(t)$ was static until period 2. Then, a new norm or a new diagnostic method arrives in period 3 e.g. an MRI or CAT scan for spotting malignant tumors, or an institutional change such as the National Insurance Scheme begins, allowing more people access to the general practitioner. People in period 3 get to know they have, or can potentially have—say, breast cancer—something they could not know until period 2, and the cancer had gone untreated until it was too late.

Broadly, one could think of this as growing awareness of symptoms—that lump in the breast, say—prompting people to seek diagnoses, and then medical remedy if recommended and if it is available. The diagnostics can be tests for blood pressure, blood-sugar levels, blood clots, cholesterol, vitamin deficiencies, fractures, concussions, among a host of others. So what looks like inflation of reported-illness rates could actually be an expression of more demand for medical care.

But what may be improper is to see growing reported illness rates as some biological issue such as increasing frailty. The advent of new technology and the discovering of illness are not purely biological matters. Further, the issue in this scenario is not one of biological underpinnings. The issue is the demand for healthcare. Inflation may well be an expression of the growing underlying demand, of wanting to find out if something is seriously wrong, of fixing it if an effective technology is at hand, and of grabbing the opportunity when a new technology or diagnostic procedure arrives so that the outcome of illness is not fatal. Further, growing use of technology, or more visits to the physician, does not necessarily mean people have become less healthy or frailer than before. They could just be better informed than before and are seeking expert advice. Or, they may have gained access to medical services that were unavailable until then. To make an analogy, should a time series on real consumption be seen as inflated or defective because new types of goods and services are being consumed? Why should more reported-illness rates, or the inflation of morbidity, be suspect if the underlying reasons are better diagnostics, treatments, and medical services? What should raise concern, instead, is if no such demand or if insufficient demand were being expressed in survey data, especially when death rates were growing.

If so, information on reported-illness rates would actually be very valuable. Instead of treating it with caution, scholars may want to extract from it signals on how much of the growing demand is temporary versus permanent; how much of it is for short-lived ailments versus for lasting ones; and how much of the change in trend is the result of a better diagnostic technology or norm.

Returning to the scenario, the likely result of the new diagnostic technology, or of the new awareness, is an increase in *reported* illness, changing the situation from static illness profiles to ones that show expansion of illness: period 3 onward $i(t, a)$ shifts up and becomes flatter:

$$\beta_i(1) = \beta_i(2) > \beta_i(3) \tag{10.9b}$$

Depicted in Fig. 10.3a, the profile shifts from $i(1, a) = i(2, a)$ to $i(3, a)$, so that $\beta_i(1) = \beta_i(2) > \beta_i(3)$. What is the ramification for per capita healthcare spending $h(t)$? In period 2, the one before which the diagnostic technology became available, spending can be written as:

$$h(2, a) = C_r(i(2, a)) + \Omega\,(d(2, a)), \tag{10.9c}$$

where C_r is the cost of treating illness rates in the profile. It embodies the state of treatment technology in period 2. Technological gain can shift this cost function downward, reducing $h(t)$. Moreover, all else equal, it could shift up as illness profile $i(2, a)$ shifts up (illness rates grow); would shift downward if $i(2, a)$ shifts down (illness rates fall). Likewise, Ω is a death-related cost schedule, involving intense medical treatment of the debilitating state before death as many studies cited in Chap. 9 have found. Death-related costs can occur at each age of the profile. The costs shift when the aging-profile $d(2, a)$ shifts. These two cost schedules will be used as the main arguments in spending profiles such as Eq. 10.9c.

Recall, Abel-Smith and Titmuss (1956) had faced just the situation in Eq. 10.9c. For the 1950s, they had some measure of $i(2, a)$ and $C_r(i(2, a))$ from the most recent Census. They did not know how, and why, $i(2, a)$ and $d(2, a)$ might change. Without the knowledge, they let $i(2, a)$ stay static, but more or less scaled it up by the Registrar-General's population projections to come up with $h(3, a)$, which in their study was 20 years out. Meanwhile $d(2, a)$ in their analysis did not shift as it was subsumed in the scaling ("the mortality assumptions").

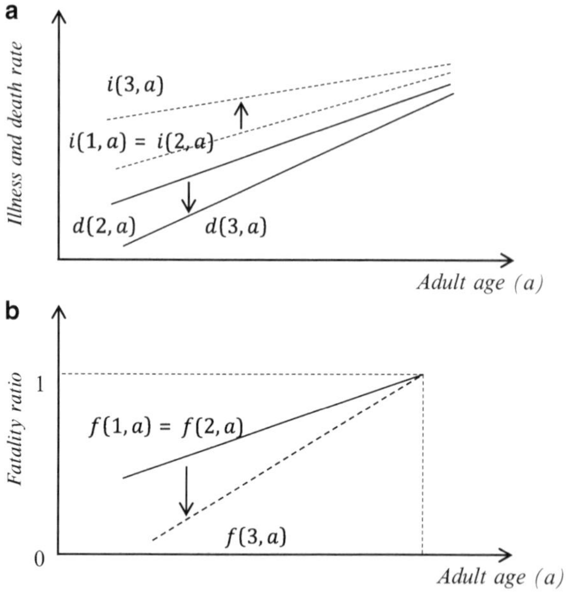

Fig. 10.3 Stylized age-profiles for expansion of illness and compression of deaths (**a**) Illness and death rates (**b**) Fatality ratios

To see how this spending profile is sloped, take the first derivative of Eq. 10.9c with respect to age (a):

$$h^{'}(2,a) = c_r\beta_i(2) + \omega\beta_d(2), \tag{10.9d}$$

which says that $h(2, a)$ slopes upward because: (1) $i(2, a)$ and $d(2, a)$ do so, or as $\beta_i(2)$ and $\beta_d(2)$ are > 0; (2) marginal treatment cost (c_r) and the marginal death-related cost (ω) are positive and rise with age (or are upward sloping). Steeper the slopes $\beta_i(2)$ and $\beta_d(2)$, all else constant, steeper the profile of the marginal costs, and thus steeper the spending profile. Further, any technology that *shifts* marginal treatment cost or marginal death-related cost, up or down, all else equal, would shift the spending profile accordingly.

Compared to the residual in Chap. 9, some of the choices for cost containment now appear explicitly in Eqs. 10.9c and 10.9d. If costs are a positive function of $i(t, a)$ and $d(t, a)$, then mechanically slashing the costs could adversely affect at least $d(t, a)$ and $\beta_d(t)$ for any given $i(t, a)$ and $\beta_i(t)$ because illnesses may go untreated. The better options seem to be: (1) shift the cost curves down for any given set of profiles, by using better technology with lower unit costs (e.g. Cutler 2014); (2) reduce the level of $i(t, a)$ and $d(t, a)$, for a given technology or cost schedule. That is, cost containment may be achieved by using effective but less costly technologies, and by reducing the illness rates and death rates to whatever extent possible given the lifespan's limit.

Continuing with the scenario, in period 3, people are more aware and seek diagnoses supplied at some cost C_D. They are then prescribed medical treatment, which is supplied at cost C_r. The success rate of the treatment, an exogenous parameter of the technology in use, should sway $\beta_d(t)$. Denote the success rate as $\theta \in (0, 1)$. For θ closer to 0, the death rate associated with any $i(t, a)$ is very low, and $d(2, a) > d(3, a) = \theta\, i(3, a)$. I will maintain $d(3, a)$ as the relevant term below to relate the outcomes to aging, invoking θ only where necessary.

As the compression of deaths continues for a given favorable θ, $\beta_d(3)$ would increase; hence the shift to $d(3, a)$ in Fig. 10.3a. However, $\beta_i(t)$ had shrunk (shifted up to $i(3, a)$). With greater illness rates but lesser death rates, the fatality ratio's profile shifts to $f(3, a)$ in Fig. 10.3b.

Per capita spending in period 3 can be written as:

$$h(3,a) = C_D(i(3,a)) + C_r(i(3,a)) + \Omega(d(3,a)). \tag{10.9e}$$

The new elements in this profile are C_D, and the period 3 profiles of illness and aging. The values in this equation were what Abel-Smith and Titmuss (1956) likely wanted to know. However, it has diagnostic technology that lets people find out if medical treatment is required, which relaxes their assumption of the unchanged "incidence and character of sickness and injury." One could introduce another parameter to indicate that not all diagnoses require further medical treatment (by introducing a constant fraction in the second term), but its absence does not change the main result. Taking the first derivative of Eq. 10.9e with respect to age (a):

$$h'(3,a) = (c_D + c_r)\beta_i(3) + \omega\, \beta_d(3), \tag{10.9f}$$

which slopes upward for reasons outlined earlier. What is the change in $h(t)$ between period 2 and 3? Subtract Eq. 10.9c from Eq. 10.9e:

$$\Delta h(3,a) = C_D \ (i(3,a)) + C_r(\Delta i(3)) + \Omega\,(\Delta d(3)), \tag{10.9g}$$

where, change in illness rates $\Delta i(3) = i(3,a) - i(2,a) > 0$, but the change in death rates $\Delta d(3) = d(3,a) - d(2,a) < 0$.

Say for the moment that illness In Eq. 10.9g were not an issue whatsoever and $i(3,a) = i(2,a) = 0$; no diagnostics costs were ever incurred ($C_D = 0$); and there were no illness requiring medical treatment ($C_r = 0$). The change in spending then would be:

$$\Delta h(3,a) = \Omega \Delta d(3) < 0 \tag{10.9h}$$

since $\Delta d\,(3) < 0$. When there are only delays in deaths, healthcare spending shrinks for any given Ω. "Pure" aging without illness would not need as much healthcare spending, and would actually reduce it. This explains:

> When age-specific death rates fall over time, there are fewer people in the last year of life at any age; thus their expenditures may be less than those predicted from a previous age-spending pattern. (Fuchs 1990, p. 536)

But clearly that is not the complete story. With reported illness expanding ($\Delta i(3, a) > 0$), resources devoted to medical treatment expand, and healthcare spending grows. The elements that boost it are: (1) diagnostic costs ($C_D\, i(3, a)$); (2) medical treatment costs (C_r); and (3) growth of illness rates ($\Delta i(3) > 0$).

One may also express the issue by subtracting Eq. 10.9d from Eq. 10.9f to see how the slope of the spending profile changes:

$$h^{'}(3) - h^{'}(2) = c_D \beta_i(3) + c_r \Delta\beta_i(3) + \omega \Delta\beta_d(3), \qquad (10.9\text{i})$$

where $\Delta\beta_i(3) = \beta_i(3) - \beta_i(2) < 0$ (illness rates grow and the profile become flatter), $\Delta\beta_d(3) = \beta_d(3) - \beta_d(2) > 0$ (death rates fall and the profile become steeper). The slopes of the two profiles tug the spending profile's slope in opposing directions. Where it ends up depends on which one tugs harder.

Equation 10.9f thus conveys that following elements can shape the slope of the profile of $h(t)$: (1) marginal diagnostic cost c_D; (2) slope of the current illness rate profile $\beta_i(3)$ or the current demand for diagnoses; (3) marginal treatment cost c_r; (4) change in the slope of the illness profiles $\Delta\beta_i$, which is positive if illness rates fall and is negative if they grow; (5) marginal death-related cost ω; (6) change in aging profile's slope $\Delta\beta_d$, which is positive when death-rates fall and negative when they rise. All else equal, the combination of shrinking $\beta_i(t)$ (more illness) and growing $\beta_d(t)$ (more aging) can boost spending.

If there were imperfect competition in the supply of technology, there would be a markup above marginal costs. This would result in: (1) higher *level* of the spending profile; (2) under-utilized resources that come with market imperfections such as patents and monopoly rights; (3) if social benefits are larger than the private ones to the suppliers of technology, there may be underinvestment in fundamental R&D required to keep pace with the diverse non-communicable diseases causing illness; and (4) there could be welfare loss in terms of illnesses going untreated (because of underinvestment in discovery), leading possibly to greater death rates than would otherwise be the case. Excepting the last one, these are standard arguments derived in models of endogenous technological change (Romer 1990; Aghion and Howitt 1992). Furthermore, as Weisbrod (1991) has argued, technological change in the medical sector seems to have had a symbiotic relation with the scope of healthcare insurance. While the expansion of insurance may help pay for cost-increasing technologies (thus encouraging bio-medical technological change), technological change itself could define the scope of illnesses that can be insured. Thus, both interdependently raise expenses per capita at the aggregate level. Under the current framework, if the diagnostic technology is seen as an exogenous introduction of insurance that lets people discover any health problems, then the trends in the costs of treatment technologies, all else equal, may be the ones that determine the path of healthcare spending after the new insurance regime is in place. Assuming diagnostic costs are less than treatment costs, then a downward shift in the treatment cost schedule, for a given illness profile, may also help slow spending growth. Further, though the diagnostic technology serves to increase spending, not having it could result in illness going untreated (welfare loss) and that may have an adverse effect on aging (as more deaths result). I will return to this issue a little later.

Meanwhile, the measure of aging $\beta_d(t)$ had yielded a positive coefficient in Chap. 9. Equation 10.9h suggests that $\beta_d(t)$ must: (1) co-vary inversely with $\beta_i(t)$ (more illness rates and thus more spending); (2) co-vary positively with diagnostic and treatment costs; and (3) co-vary positively with death-related costs. If so, based on this framework, an interpretation of the results would be that reported illness rates may have grown as people became more aware and sought more medical attention to prevent fatality, which they likely achieved, at least through the mid-1980s. With more reported illnesses, treatment costs grew, and spending climbed as $\beta_d(t)$ had progressed in the age of delayed deaths.

For cost containment, the result makes the ramifications for wellbeing clearer: if $\beta_i(t)$ and $\beta_d(t)$ co-vary inversely, then slashing a given c_r or c_D to slow Δh may only shrink $\beta_d(t)$—less aging—as some of the illnesses go untreated or unattended. A welfare-improving option instead is to shift the cost schedules down by lowering unit costs that come from more effective technologies that reduce θ.

Within this setup one could ask: what happens to healthcare spending if people were unaware of the type of illness ($C_D = 0$), had some minimal knowledge or technology to remedy growing illness rates ($C_r \approx 0$, $\theta \approx 1$), and had few ways or none to prolong the lives of people who were (in retrospect) terminally ill ($\Omega \approx 0$). Suppressing a, rewrite Eq. 10.9g more generally as:

$$\Delta h(t, a) = C_D \ i(t) + C_r \ \Delta i(t) + \Omega \ \Delta d(t), \tag{10.9j}$$

There would be less spending, since on the right-hand-side: (1) the first term is zero without diagnostic technology; (2) $\Delta i(t) > 0$, but with $C_r \approx 0$, the effective technology is inaccessible to most or may not exist, and so this term is very small; and (3) without the technology's benefit to aging ($\theta \approx 1$), aging reverses, making $\Delta d(t) > 0$, even as $\Omega \approx 0$. This can be seen in terms of the changing slope of the spending profile as well. Rewrite Eq. 10.9i generally as:

$$h'(t) - h'(t-1) = c_D \beta_i(t) + c_r \Delta\beta_i(t) + \omega \Delta\beta_d(t), \tag{10.9k}$$

where $\Delta\beta_i(t) < 0$ because of growing illness rates and flatter profile and $c_r \approx 0$; and with $\theta \approx 1$, $\Delta\beta_d(t) < 0$ as death rates across profiles grow, and the slopes become flatter while $\omega \approx 0$. All of them convey that the spending profile either remains unchanged despite the adverse situation, or it becomes flatter. These conditions for a flatter profile do not seem to be healthy aging.

Equations 10.9j and 10.9k may explain the conditions in the second half of the nineteenth century. Without the technology, spending is likely to have been minimal even as illness rates potentially grew, and the welfare consequence was likely adverse as death rates climbed.

If spending on diseases that made up nearly half the total disease-caused deaths (Chap. 2) was presumably low back then, the advent of newer norms, diagnoses, and technologies may have been crucial to the growth of spending since. The norms can be social, medical, and political. The social norms may be in the general awareness of symptoms, in the threshold of discomfort or pain at which to seek medical help;

and in institutions such as the National Insurance Scheme that enabled wider access to general practitioners. The medical norms may have been technological or knowledge-based, such as how to spot illness or diagnose it, and what type of treatment to prescribe. When diseases become treatable, or curable, the medical expense for the treatment may increase, prompting demand for medical insurance which pools risks. If the medical treatment significantly enhances the quality of life, it may come to be viewed as an "altruistic externality" and may lead to a more fundamental political norm, such as a change of regime from imperfect health insurance to universal insurance (Weisbrod 1991; Pauly 1986). Each type of norm may contribute to the 'inflation' of illness rates. However, it may be unsafe to label it that way, a priori, because what may seem like inflation (a defect requiring correction) could well be the correct signal of the underlying demand for remedying illness.

However, this scenario does not exhaust the set of possibilities. It helps to explore a couple more, if only to identify ones that may not be capable of generating growth of healthcare spending.

10.4 As Does Prevention, but Differently

As death rates fell over the twentieth century, a central concern since very early in the debate had been whether illness rates were falling, especially among the middle-to-older age people, and what could be done if illness rates were not falling (e.g. Fries 1980, 1989, 2005; Manton et al. 1982, 1997; Fogel 1999; Steckel and Costa 1997). If they were falling, the question was: why per capita healthcare spending was growing? The previous section stylized one scenario for its growth. Another question that can be asked is: what would be the profile of $h(t)$ if policy prioritized the prevention of illnesses that accompany aging? Analyzing this type of situation serves as an input to scenarios developed in subsequent sections.

It is best to begin with scenario II. Period 1 now is in scenario II where compression of deaths has been occurring since the past for various reasons, but illness rates have been static. Per capita healthcare spending is:

$$h(1,a) = C_r(i(1,a)) + \Omega(d(1,a)), \tag{10.10a}$$

and the slope of the spending profile is:

$$h'(1,a) = c_r \beta_i(1) + \omega \beta_d(1), \tag{10.10b}$$

In period 2, policy sees that $i(2,a) = i(1,a)$ and decides it wants to lower it by period 3. With a target $i^T(3,a) < i(2,a)$ in mind, it begins a prevention effort that costs:

$$C_p = C_p\left(i^T(3,a)\right) \tag{10.10c}$$

so that more ambitious the target $i^T(3,a)$, costlier the prevention. The first derivative of Eq. 10.10c with respect to a is:

$$C_p{}'\left(i^T(3,a)\right) = c_p\beta_i^T(3) > 0, \tag{10.10d}$$

where, all else equal, the marginal cost $c_p\beta_i^T(3)$ is higher, steeper the illness profile the policy targets for period 3. That is, it is increasingly difficult to prevent illness rates as one goes further up the age-profile to older age groups. However, to achieve the results, expenses are incurred in period 2. The spending profile in period 2:

$$h(2,a) = C_p\left(i^T(3,a)\right) + C_r(\,i(2,a)) + \Omega(d(2,a)), \tag{10.10e}$$

and its slope,

$$h'(2,a) = c_p\beta_i^T(3) + c_r\beta_i(2) + \omega\beta_d(2), \tag{10.10f}$$

and the change in per capita spending between period 1 and 2 is Eq. 10.10e minus Eq. 10.10a:

$$\Delta h(2,a) = C_p\left(i^T(3,a)\right) + \Omega(\Delta d(2)), \tag{10.9g}$$

since $\Delta i\ (2) = 0$ because of static illness rates between the first two periods. The spending grew because of prevention costs, but fell because of the lower cost of "pure" aging in the compression as $\Delta d\ (2) < 0$. Their net effect sets the outcome. Meanwhile, change in the slopes of the profiles, Eq. 10.10f minus Eq. 10.10b, is:

$$h'(2,a) - h'(1,a) = c_p\beta_i^T + \omega\,\Delta\beta_d(2). \tag{10.10h}$$

Since $\Delta\beta_i\ (2) = 0$ (static illness rates), and as both $c_p\beta_i^T > 0$ and $\Delta\beta_d(2) > 0$, the profile becomes steeper.

The $h(t)$ in period 3 is:

$$h(3,a) = C_r\left(i(3,a) - i^T(3,a)\right) + \Omega(d(3,a)), \tag{10.10i}$$

which depends on how far the targeted $i^T(3,a)$ was able to reduce $i(3,a)$. If $i(3,a) - i^T(3,a) = \Delta^T i = 0$, then there is no illness left and death-related costs are the only expense left in Eq. 10.10i. More realistically, if $i(3,a) - i^T(3,a) > 0$, policy succeeds if $i(3,a) - i^T(3,a) < i(2,a)$. The slope of the spending profile is:

$$h'(3,a) = c_r(\beta_i(3) - \beta_i^T(3)) + \omega\beta_d(3), \quad (10.10j)$$

The change in *h(t)* is Eq. 10.10i minus Eq. 10.10e, suppressing *a* for convenience:

$$\Delta h(3) = C_r(i(3) - i^T(3)) - C_p(i^T(3)) + \Omega(\Delta d(3)). \quad (10.10k)$$

If $i(3) - i^T(3) < i(2), C_r$ will diminish relative to period 2 since the illness profile has shifted down; and as $C_p(i^T(3)) > 0$ and $\Delta d(3) < 0$, the outcome is $\Delta h(3) < 0$. Or, spending falls. This reduction in spending hinges on the condition $i(3) - i^T(3) < i(2)$.

The change in the slope is Eq. 10.10j minus 10.10h or $h'(3) - h'(2)$

$$\Delta h'(3) = c_r\, \Delta\beta_i(3) + (c_r - c_p)\beta_i^T(3) + \omega\, \Delta\beta_d(3). \quad (10.10l)$$

Here $\Delta\beta_i(3) > 0$ and $\Delta\beta_d(3) > 0$ make the new profile steeper. Whereas, only if $c_r < c_p$, is there some tendency for the profile to become less steep: if the marginal cost of medical treatment for the new (lesser) illness rates in period 3 turns out less than the marginal prevention costs incurred in period 2, the spending profile would become flatter than it would otherwise have been. This is a conditional outcome.

10.5 Pure Compression Shrinks Spending

To consider another scenario, say there were neither prevention nor norms influencing illness rates and death rates, and both were compressing on their own without any use of diagnostic technology or prevention. The *h(t)* in the two periods would be:

$$h(2,a) = C_r(i(2,a)) + \Omega(d(2,a)) \quad (10.10m)$$
$$h(3,a) = C_r(i(3,a)) + \Omega(d(3,a)) \quad (10.10n)$$

and the change in the spending:

$$\Delta h(3) = C_r(i(3,\mathrm{a}) - i(2,a)) + \Omega(d(3,a) - d(2,a)), \quad (10.10o)$$

or more generally,

$$\Delta h(t) = C_r(\Delta i(t,\mathrm{a})) + \Omega(\Delta d(t,a)). \quad (10.10p)$$

The medical treatment costs fall as $\Delta i(t, a) < 0$ and the death related costs shrink as $\Delta d(t, a) < 0$. Or, healthcare spending falls as both illness rates and death rates fall, for a given technology. All else equal, containing costs (to slow healthcare spending) is a non-issue.

To see the slopes of the profiles in this outcome, take first derivatives of Eqs. 10.10m and 10.10n and then subtract one from the other, and express them in general terms:

$$\Delta h'(t) = c_r \Delta\beta_i(t) + \omega\Delta\beta_d(t), \tag{10.10q}$$

where, since $\Delta\beta_i(t) > 0$ and $\Delta\beta_d(t) > 0$, the spending profile keeps shifting down and becoming steeper, coinciding with less healthcare spending. All else equal, compressing illness and death rates co-vary positively. 'Healthy aging' thus has steeper profiles of illness rates, of death rates, and of spending. This scenario, by itself, has likely not been the historical experience for healthcare spending. However, it can play a role within a larger scenario explored in the next section.

Meanwhile, how does this case differ from the one with diagnostic costs? For a given technology, death rates and illness rates co-vary positively in this scenario, but co-varied negatively in the one with diagnostic technology. The norm-driven scenario led to growth of spending, but this one led to reduced spending.

Table 10.2 summarizes the various outcomes. In scenario C of pure compression or 'healthy aging' just discussed, all profiles become increasingly steeper and spending falls over time. In scenario B, where illness and death rates were growing, technology is minimal, the profile slopes become flatter, and spending falls. Only in scenario A does healthcare spending grow. Further, scenario D is special. It is the only one that links spending between periods, an intertemporal link from prevention in one period to illness rates in the next. It generated a transition from static illness rates to compressing illness rates. That transition, however, led to less per capita healthcare spending, but conditionally. Table 10.2 suggests that the long-term *growth* of spending, combined with the transition of illness rates and death rates may be derivable by combining scenario A and D, the subject of the next section.

10.6 Intertemporal and Life-Course Channels Show the Way

New elements need to be brought in to explore that combination. Recall, each $\beta_d(t)$ was made up of a childhood-linked $\beta_d^C(t)$ and a life-course $\beta_d^L(t)$. So it must be that each aging profile $d(t, a)$ has death rates that are swayed by developments in the past, denoted $d^C(t, a)$ and by current events denoted $d^L(t, a)$. Similarly, $i(t, a)$ has $i^C(t, a)$, $i^L(t, a)$, $\beta_i^C(t)$ and $\beta_i^L(t)$. For convenience, the notation hereon will suppress a, roping it in only where necessary.

To start with a basic case, say both childhood and life-course elements play a role in each period. In the adopted format, five cohorts contribute to each $\beta_d^C(t)$. It took change over five birth-decades to sway each $\beta_d^C(t)$ in the series. Further each cohort made its contribution to five $\beta_d^C(t)$, each time at a different age. Instead of writing in terms of five contributions per period-profile, it is less unwieldy to think of period 1 and period 2 as being decades apart and express the essential idea in terms of just one $\beta_d^C(t)$.

Table 10.2 Scenarios of per capita healthcare spending growth in different aging transitions

	Scenario	$\Delta\beta_i$(t) and $\Delta\beta_d$(t)	Change in the slope of spending profile $\Delta h'(t)$	Growth: Δh(t)
A	Norms: C_D, C_r, $\Omega > 0$, $\Delta i(t) > 0$, $\Delta d(t) < 0$	$\Delta\beta_i(t) < 0$, $\Delta\beta_d(t) > 0$	>0	>0
B	No technology: C_D, C_r, $\Omega \approx 0$, $\Delta i(t) > 0$, $\Delta d(t) > 0$	$\Delta\beta_i(t) < 0$, $\Delta\beta_d(t) < 0$	<0	<0
C	Pure compression: $C_D = 0$, C_r, $\Omega > 0$, $\Delta i(t) < 0$, $\Delta d(t) < 0$	$\Delta\beta_i(t) > 0$, $\Delta\beta_d(t) > 0$	>0	<0
D	Prevention: Cp, $\Omega > 0$, $\Delta i(t) < 0$ conditional, $\Delta d(t) < 0$	$\Delta\beta_i(t) > 0$, $\Delta\beta_d(t) > 0$ Conditional	>0 Conditional	<0 Conditional

In period 1, healthcare expense is:

$$h(1) = C_r\left(i^C(1) + i^L(1)\right) + \Omega\left(d^C(1) + d^L(1)\right). \tag{10.11a}$$

To simplify matters, say treatment costs and death-related costs apply to both the life-course and the childhood-linked components. That is, the cost schedules reflect the state of technology in use system-wide, and not the relative shares of the cost of treating childhood-linked and life-course components. Likewise, in period 2:

$$h(2) = C_r\left(i^C(2) + i^L(2)\right) + \Omega\left(d^C(2) + d^L(2)\right), \tag{10.11b}$$

and subtracting Eq. 10.11a from 10.11b yields

$$\Delta h(2) = C_r\left(\Delta i^C(2) + \Delta i^L(2)\right) + \Omega\left(\Delta d^C(2) + \Delta d^L(2)\right), \tag{10.11c}$$

and similarly, the change of slope $h^{'}(2) - h^{'}(1)$, or $\Delta^{'} h(2)$:

$$= c_r\left(\Delta\beta_i^C(2) + \Delta\beta_i^L(2)\right) + \omega\left(\Delta\beta_d^C(2) + \Delta\beta_d^L(2)\right) \tag{10.11d}$$

If there were no change in the childhood regimes, and everything was life-course-driven, Eq. 10.11c, expressed generally, reduces to:

$$\Delta h(t) = C_r\left(\Delta i^L(t)\right) + \Omega\left(\Delta d^L(t)\right) \tag{10.11e}$$

and Eq. 10.11d to

$$\Delta h^{'}(t) = c_r\left(\beta_i^L(t)\right) + \omega\left(\beta_d^L(t)\right), \tag{10.11f}$$

both providing a partial picture. Without the superscript L, they may as well be one of the scenarios explored earlier. So this situation alone is unsatisfactory. To see why change in both childhood-linked and life-course developments should matter individually, ask what would happen to spending if neither component changes. Equation 10.11d says that there should be no change in spending. One component or the other in this setup must change to alter spending.

Say policy wants to shrink the illness profiles in some future period. Since it cannot now foresee the life-course elements, it can only sway the childhood-linked elements of a future illness profile. This requires an intertemporal connection between the current childhood regime and future $i^C(t, a)$, a feature like the prevention cost in scenario D.

The long-reach of prevention show up only through $i^C(t, a)$ and $\beta_i^C(t)$. The life-course elements, by definition, have no such link to the past. The assumption is that there are no interaction terms. As was discussed in Chap. 9, it does not necessarily provide fresh insight, but only complicates the algebra by including a third term with superscript 'CL' in each profile.

Write the resources devoted, or costs incurred, for prevention in period 1 as

$$C_p(1,a) = C_p\left(i^C(2,a)\right) \tag{10.12g}$$

where i^C (2, a) is the childhood-linked illness profile in period 2 that can result from prevention in period 1. Suppose $i^C(2, a) < i^C(1, a)$ for now. It can be changed later to study what happens when the dynamic is thrown into reverse due to adverse developments in childhood. Take the first derivative of Eq. 10.12g:

$$C_p^{'}(1,a) = c_p\beta_i^C(2,a). \tag{10.12h}$$

For a given prevention technology, its marginal cost rises with $\beta_i^C(2) > \beta_i^C(1)$. Think of c_p as some technology or new knowledge, like the ascendancy of the germ theory of diseases in the nineteenth century. It could also be the efficacy of the resources devoted to public health infrastructure then. It could be better health of mothers during pregnancy. Structurally, it could mean reduction of political or economic inequality, to the extent they may relate to future illness rates. But importantly, no matter what it is, c_p must involve children's physiological development. All profiles are defined over the adult stages, and their only link to the past is through what happened in childhood. With the expenses incurred in period 1, the spending profile is:

$$h(1) = C_p\left(i^C(2)\right) + C_r\left(i^C(1) + i^L(1)\right) + \Omega\left(d^C(1) + d^L(1)\right) \tag{10.12i}$$

and its slope:

$$h^{'}(1) = c_p\beta_i^C(2) + c_r\left(\beta_i^C(1) + \beta_i^L(1)\right) + \omega\left(\beta_d^C(1) + \beta_d^L(1)\right) \tag{10.12j}$$

Meanwhile, $h(2)$ is the same as Eq. 10.11b:

$$h(2) = C_r\left(i^C(2) + i^L(2)\right) + \Omega\left(d^C(2) + d^L(2)\right), \tag{10.12k}$$

and its slope

$$h^{'}(2) = c_r\left(\beta_i^C(2) + \beta_i^L(2)\right) + \omega\left(\beta_d^C(2) + \beta_d^L(2)\right), \tag{10.12l}$$

The $\Delta h(t)$ between two periods can be written generally as:

$$\Delta h(t) = C_r\left(\Delta i^C(t) + \Delta i^L(t)\right) + \Omega\left(\Delta d^C(t) + \Delta d^L(2t)\right) - C_p i^C(t) \tag{10.12m}$$

Suppose the life-course issues do not matter and only the childhood-linked components sway the profiles. Ridding Eq. 10.12m of all the life-course terms yields:

$$\Delta h(t) = C_r\,\Delta i^C(t) + \Omega\,\Delta d^C(t) - C_p i^C(t) < 0 \tag{10.12n}$$

That is, there is compression of illness rates, $\Delta i^C(t) < 0$, and of death rates, $\Delta d^C < 0$. Since $i^C(2,a) < i^C(1,a)$, it follows that $\theta\, i^C(2,a) < \theta\, i^C(1,a)$ or $d^C(2,a) < d^C(1,a)$ for $\theta \in (0, 1)$.

Some of the drop in $\Delta h(t)$ is clearly due to more spending in the earlier period. But the more important issue is that previous period's spending fostered new illness rate profiles that lowered this period's spending. If there were several periods over which prevention costs were incurred, the time series of healthcare spending would grow by the amount of the prevention cost in each period. But in each period, it would also shrink by the prevention cost in the previous period combined with lesser expenses required to treat the reduced illness rates and lesser death-related costs. For instance, if another prevention cost was incurred in period 3, then it must more than offset all the elements in Eq. 10.12n for spending to keep growing. Since the additional spending required lessens in each new period (as illness and death rates fall), spending growth slows. The only way this scenario can lead to more growth is if cohorts of children had fared worse. Instead of being prevention, C_p stood for something adverse and childhood-related illness and death rates grew. If not, then something else—life-course elements—must have boosted $h(t)$ over time.

The diagnostic technology, interpreted broadly, can help bring in the life-course elements. The diagnostics are supplied at cost C_D only to the life-course components of the profiles. By introducing it, the reported life-course illness rates grow and Eq. 10.12k becomes:

$$\begin{aligned} h(2) = {} & C_D\left(i^L(2)\right) + C_r\left(i^C(2) + i^L(2)\right) \\ & + \Omega\left(d^C(2) + d^L(2)\right) \end{aligned} \tag{10.12o}$$

The change in spending, Eq. 10.12o minus Eq. 10.12i written generically is:

$$\begin{aligned} \Delta h(t) = {} & C_D\left(i^L(t)\right) - C_p\left(i^C(t)\right) + C_r\left(\Delta i^C(t) + \Delta i^L(t)\right) \\ & + \Omega\left(\Delta d^C(t) + \Delta d^L(t)\right) \end{aligned} \tag{10.12p}$$

Collecting childhood-linked and life-course elements:

$$\begin{aligned} \Delta h(t) = {} & C_r \Delta i^C(t) + \Omega \Delta d^C(t) - C_p\left(i^C(t)\right) \\ & + C_D\left(i^L(t)\right) + C_r \Delta i^L(t) + \Omega\,\Delta d^L(t) \end{aligned} \tag{10.12q}$$

If the childhood-linked elements do not matter, the outcome is:

$$\Delta h(t) = C_D i^L(t) + C_r\,\left(\Delta i^L(t)\right) + \Omega\left(\Delta d^L(t)\right) \tag{10.12r}$$

where $\Delta h(t) > 0$ only if $C_r(\Delta i^L(t)) + C_D(i^L(t)) > \Omega(\Delta d^L(t))$. Healthcare spending grows if diagnostic costs and the medical costs of the growing reported illness rates is more than the fall in death-related costs associated with shrinking death rates. If the life-course parts do not matter, the outcome becomes Eq. 10.12n, which results in less spending.

It follows that the 'transitions of aging' must involve both components (childhood and life-course) for per capita spending to have grown. What drives growth in the childhood-linked part of the story is resources devoted to prevention such as better public health infrastructure to curb infectious diseases, especially in children. It is linked intertemporally to future $i^C(t)$ and $d^C(t)$ that tended to reduce spending. If that were the only story, spending would slow because illness rates and death rates fall. For spending to keep growing, the life-course part of the story must continually stay in the picture. The norm-driven demand for diagnoses and medical treatment raise $i^L(t)$, but the new technology curtails $d^L(t)$. Both $d^C(t)$ and $d^L(t)$ fall, but only $i^C(t)$ falls with them; whereas $i^L(t)$ grows.

One may ask Eq. 10.12q: what type of transition in aging occurs when children's average height shrinks the way it did in the nineteenth century, or when public health infrastructure deteriorates, or when mothers fare poorly, or if income inequality worsens to adversely affect children's physiological development? The childhood-linked parts of the picture are thrown into reverse, and the illness and death rates grow. Meanwhile, with existing diagnostic and treatment technologies (C_D, C_r) the overall spending climbs because the childhood-linked segments that would otherwise have subdued spending, now fuel it instead.

One may also ask: what if (C_D, C_r), as embodiments of technology, become potentially inaccessible to most, or do not exist for many unyielding diseases? Now the life-course components of the picture worsen, and healthcare spending would fall. Just when the latent demand might have grown, it finds that medical remedies to spend on, either do not exist or are inaccessible. The ramifications for wellbeing are adverse.

To the issue of how death rates and illness rates relate during the Transition, the equation says that $i^C(t)$ would relate positively to $d^C(t)$ and to $d^L(t)$, but $i^L(t)$ may not relate with them positively at all times. Only when there is no effective treatment technology for non-communicable diseases, and when childhood development is poor, the trends of all of them may correlate positively—which might have been the situation in the nineteenth century.

Expressing Eq. 10.12q in terms of marginal costs and slopes:

$$\begin{aligned}\Delta h'(t) &= c_r \Delta\beta_i^C(t) + \omega\Delta\beta_d^C(t) - c_p\beta_i^C(t) \\ &\quad + c_D\beta_i^L(t) + c_r\Delta\beta_i^L(t) + \omega\Delta\beta_d^L(t)\end{aligned} \tag{10.12s}$$

For given marginal costs $c_r, \omega, c_p, c_D, \Delta\beta_i^C > 0$, but $\Delta\beta_i^L < 0$, so the slope of the overall illness profile can tilt either way depending on which component sways it more. In contrast, both $\Delta\beta_d^C > 0$ and $\Delta\beta_d^L > 0$, which means $\Delta\beta_d^C + \Delta\beta_d^L = \Delta\beta_d > 0$, and healthcare spending grows alongside aging. This connects with the results in the last chapter of higher β_d boosting healthcare spending. However, its relation to β_i may evolve, depending on whether β_i^C or β_i^L dominates. To conjecture, before the advent of diagnostic technologies β_i^C may have been heftier, but after then β_i^L may sway β_i more, but this cannot be guaranteed especially as the life-course component is likely very unpredictable.

Returning to the issue of cost containment, for a given illness profile, any technology that lowers the unit costs may slow the growth of healthcare spending by shifting the cost curves down. However, this result may not obtain if the technology is supplied under imperfect competition where monopolists charge a markup above marginal cost. Higher costs may still accompany the better medical outcomes, the unit costs could drop in quality-adjusted terms, but they do not show up that way in the spending data. For containing costs, however, the exception is c_D. Without incurring such costs, illness may not be diagnosed in a timely manner and may even go untreated. This could reduce β_d^L and hinder aging.

Moreover, for the long term, a vital element for containing costs in the equations is c_p. It enables the childhood-linked compression in $i^C(t)$ and $d^C(t)$, both of which reduce spending, or at least slow its growth.

10.7 Perhaps Even in the Ongoing Transition

Equations 10.12o, 10.12p, 10.12q, 10.12r, and 10.12s could help think about the new expansion-like trend that has been developing in the period-β_d since the mid-1980s.

The first possibility is that β_d^L have started to worsen since the 1980s. This could happen because of diseases such as many types of cancers for which perfect medical solutions are elusive. As the heft of such diseases grows, it could imply slower spending, but accompanied by terrible loss of life.

Second, the illness rates could have been increasing, and worsening β_i^L. To the extent technology has been able to diagnose medical problems better, growth in illness rates in this period may have been of the life-course type or $i^L(t)$. If $i^C(t)$ had been falling since the transition stage, then the proportion of $i^L(t)$ in overall $i(t)$ could have grown over time.

The available data on prevalence rates, however, are not organized in age-profiles to be able to assess this issue. Nonetheless, it helps to look at some prevalence rates over 1981/1982–1991/1992 when the expansion appears to have begun. The prevalence rates in Table 10.3 are based on patients consulting their general practitioners (GP) in the national GP-based morbidity survey conducted every 10 years. Nearly all categories indicate that $i^L(t)$ may have grown. As discussed earlier, reported-illness rates can be noisy and the challenge is to isolate heftier signals (serious illness) from the minor ones (short-lived illnesses). Both the serious and the intermediate illnesses grew during this time. But one still cannot conclusively say that all of the prevalence growth was $i^L(t)$, and not partly an

Table 10.3 Annual prevalence rates per 10,000 person-years at risk

Category	1981–1982	1991–1992	% Change
All diseases	7,116	7,803	10
Serious illnesses	1,439	1,829	27
Intermediate illnesses	4,160	4,741	14
Trivial conditions	5,702	6,576	15
Circulatory system	850	931	10
Nervous system	1,409	1,732	23
Digestive system[a]	718	875	23
Neoplasms	135	239	77
Respiratory system	2,696	3,070	14

Source: Fraser et al. (1997)
[a]The rates for digestive system are average of males and female rates; all others are calculated for persons

adverse $i^C(t)$ from the inter-war cohorts, who were the major contributors to the profiles during this time. If it is the latter, then the 'expansion' in this episode may well prove transitory as the childhood-linked profiles of subsequent post-transition cohorts may have improved.

Meanwhile, spending growth slowed since about the mid-1980s. The slower growth has coincided with worsening period-β_d and growing illness rates. That could mean one of two things: (1) cost-containment (C_r) may have occurred while illness profiles may have been worsening, leading to an adverse consequence for period-β_d; (2) the types of diseases dominating the illness rates are ones that have few or no medical solutions so that for a given C_r, illness and death rate profiles shifted upward, and spending has slowed as technology for such diseases does not exist or the seriousness of the illnesses have defied medical treatment. As discussed earlier, both these scenarios have adverse ramifications for wellbeing.

Furthermore, whether the trend in this expansion continues depends partly on how the new cohorts may fare in the future, for they will shape the future period-β_d. However, it is nearly impossible to foresee their life-course elements. What one may explore instead is how the childhood-linked segments may fare as children from the newer cohorts after World War II will be the ones contributing to future period-profiles.

Figure 10.4 shows the initial states or the period-$\alpha_d(t)$s corresponding to the annual period-$\beta_d(t)$s over 1950–2000. Falling $\beta_d(t)$s and growing $\alpha_d(t)$s are not a good sign from the perspective of aging in the decades ahead. Recall from Chap. 4 that trends of the period-$\alpha_d(t)$s provide one of the first signals that something may have been amiss about the newer cohorts. The developing trend in the period-$\alpha_d(t)$s in Fig. 10.4 suggests that the new trend of the period-β_d may not be solely about older people, but also about children just entering adulthood.

It draws attention to C_p and c_p, the potential sources of intertemporal links to future period-profiles. Based on the findings in previous chapters, the reasons for rising $\alpha_d(t)$s could be: (1) infections among children; (2) stagnant or falling average

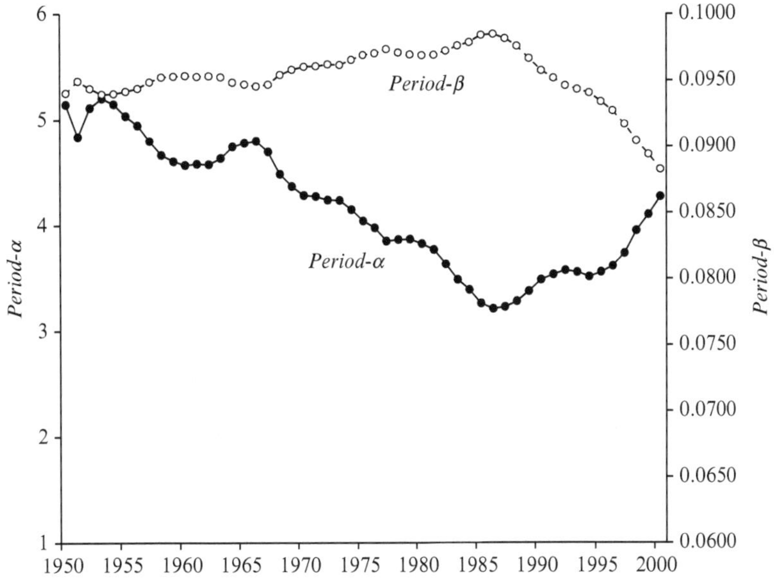

Fig. 10.4 $\alpha_d(t)$ and $\beta_d(t)$ of period-profiles, 1950–2000 (Source: author's calculations)

stature; (3) poorer health of mothers at the time of pregnancy; (4) and growing income and wealth inequality. For broader childhood development, issues such as obesity may have become critical as well.

Moreover, recent trends of both parameters seem to indicate that the age of delaying deaths has either paused or has ended. That appears to go against the view that increasingly more people can expect to live longer in the future. However, both views are compatible. The findings in the micro-level studies on middle-to-older aged people may relate more to $i^C(t), \beta_i^C$ and β_d^C (or to the interaction terms) of the profiles at older ages, while $i^L(t), \beta_i^L$ and β_d^L parts may have dominated the period-profiles explored here.

Further, this expansion is occurring in a vastly different context than the one in the second half of the nineteenth century. Nonetheless, not having technological solutions amounts to a situation similar to the late nineteenth century, and the outcomes could have the same tenor, though at an advanced economic status. Further, like then, period-life expectancies recently may have been becoming less reliable indicators of aging. The expectancies measured at advanced ages may be relevant to smaller fractions of the population associated with $i^C(t), \beta_i^C$ and β_d^C. In other words, at very older ages, childhood-linked segments of the profiles can become more prominent than life-course ones. At younger ages, however, it is the adverse life-course segments that may have dominated. This seems the only way to reconcile worsening overall period-β_d with the findings of recent micro-level research as well as with the growing ranks of very old people (Thatcher 1992, 1997).

This combination also fosters slower healthcare spending. Adverse β_i^L could be associated with diseases that defy technology. In this unfortunate scenario, more deaths result, but that slows healthcare spending (on treatment costs). The occurrence of better β_i^C too is linked to slower spending. This mix of outcomes—compression of β_i^Candβ_d^C, along with expansion of β_i^Landβ_d^L without technology—seems the only way a *globally* flatter spending profile would coincide with healthy longevity in older-aged people. An empirical exploration of whether this particular scenario has been playing out is beyond the scope of the book.

References

Abel-Smith, B., & Titmuss, R. M. (1956). *The cost of the national health service in England and Wales*. London: Cambridge University Press.

Aghion, P., & Howitt, P. (1992). A model of growth through creative destruction. *Econometrica, 60*(March), 325–351.

Akerlof, G. (1970). The market for 'lemons': Qualitative uncertainty and the market mechanism. *The Quarterly Journal of Economics, 84*(3), 488–500.

Alter, G., & Riley, J. (1989). Frailty, sickness, and death: Models of morbidity and mortality in historical populations. *Population Studies, 43*, 25–45.

Arrow, K. (1963). Uncertainty and welfare economics of medical care. *American Economic Review, 53*, 941–973.

Batty, D. G., Shipley, M. J., Gunnel, D., Huxley, R., et al. (2009). Height, wealth, and health: An overview with new data from three longitudinal studies. *Economics and Human Biology, 7*, 137–152.

Costa, D. L. (2002). Changing chronic disease rates and the long-term declines in functional limitations among older men. *Demography, 39*(1), 119–137.

Costa, D. L. (2003). Understanding the mid-life and older-age mortality declines: Evidence from the Union Army veterans. *Journal of Econometrics, 112*(1), 175–192.

Crimmins, E. M., Saito, Y., & Ingegneri, D. (1989). Changes in life expectancy and disability-free life expectancy in the United States. *Population and Development Review, 15*, 235–236.

Crimmins, E. M., Hayward, M., & Saito, Y. (1994). Changing mortality and morbidity rates and the health status and life expectancy of the older population. *Demography, 31*, 159–175.

Crimmins, E. M., Hayward, M., & Saito, Y. (1996). Mixed trends in population health among older adults. *Journal of Gerontology, 51B*, S223–225.

Cutler, D. L. (2014). *Quality cure*. Berkeley and Los Angeles: University of California Press.

Edwards, C., Gorsky, M., Harris, B., & Hind, P. R. A. (2003). Sickness, insurance and health: Assessing trends in morbidity through friendly society records. *Annales de Demographie Historique, 1*, 131–167.

Fogel, R. W. (1999). Catching up with the economy. *American Economic Review, 80*, 1–21.

Fraser, P., Fleming, D., Murphy, M., Charlton, J., Leceister, G., & Goldacre, M. (1997). Morbidity statistics from health service utilization. In J. Charlton & M. Murphy (Eds.), *The health of adult Britain 1841–1994*, HMSO, London, UK.

Fries, J. F. (1980). Aging, natural death and compression of morbidity. *New England Journal of Medicine, 303*(3), 130–135.

Fries, J. F. (1989). The compression of morbidity: Near or far? *Milbank Memorial Quarterly, 67* (2), 208–232.

Fries, J. F. (2005). The compression of morbidity. *Milbank Quarterly, 83*(4), 801–823.

Fuchs, V. R. (1990). The health sector's share of the gross national product. *Science, 247*, 534–538.

Gorsky, M., Harris, B., & Hinde, P. R. A. (2004). Health and sickness in the late nineteenth and twentieth centuries: The Hampshire Friendly Society and its records: Presented at the Economic History Society Conference, Royal Holloway and Bedford New College, London.

Gorsky, M., Guntupalli, A., Harris, B., & Hinde, A. (2011). The 'cultural inflation of morbidity' during the English mortality decline: A new look. *Social Science and Medicine, 73*(12), 1775–1783.

Gruenberg, E. (1977). The failures of success. *Milbank Memorial Fund Quarterly/Health and Society, 55*(1), 3–24.

Harris, B., Hinde, A., & Gorsky, M. (2009). The health and morbidity of Friendly Society members in the late-nineteenth century: Full Research Report ESRC End of Reward Report. Swindon: ESRC.

Harris, B., Gorsky, M., Guntupalli, A., & Hinde, A. (2011). Ageing, sickness and health in England and Wales during the mortality transition. *Social History of Medicine, 24*(3), 643–665.

Harris, B., Gorsky, M., Guntupalli, A., & Hinde, A. (2012). Long-term changes in sickness and health: Further evidence from the Hampshire Friendly Society. *Economic History Review, 65* (2), 719–745.

Johansson, S. (1991). The health transition: The cultural inflation of morbidity during the decline of mortality. *Health Transition Review, 1*, 39–65.

Johansson, S. (1992). Measuring the cultural inflation of morbidity during the decline in mortality. *Health Transition Review, 2*(1), 77–87.

Manton, K. G., Parrick, C., & Stallard, E. (1980). Mortality model based on delays in progression of chronic diseases: Alternative to cause elimination model. *Public Health Reports, 95*(6), 580–588.

Manton, K. G., Parrick, C., & Stallard, E. (1982). Changing concepts of morbidity and mortality in the elderly population. *Milbank Memorial Fund Quarterly, 60*, 183–244.

Manton, K. G., Parrick, C., & Stallard, E. (1997). Chronic disability trends in elderly United States populations: 1982–1994. *Proceedings of the National Academy of Sciences, 94*(6), 2593–2598.

Murray, J. (2003). Social insurance claims and morbidity estimates: Sickness or absence? *Social History of Medicine, 16*, 225–245.

Murray, C. L. J., & Lopez, A. D. (1996). *The global burden of disease*. Harvard: Harvard University Press.

Olshansky, S. J., Rudberg, A., Carnes, B. A., et al. (1991). Trading off longer life for worsening health. *Journal of Aging and Health, 3*, 194–216.

Pauly, M. V. (1986). Taxation, health insurance, and market failure in the medical economy. *Journal of Economic Literature, 24*(2), 629–675.

Reynolds, S. L., Crimmins, E. M., & Saito, Y. (1998). Cohort differences in disability and disease presence. *The Gerontologist, 38*(5), 578–590.

Riley, J. (1987). Ill health during the English mortality decline: The friendly societies' experience. *Bulletin of the History of Medicine, 61*, 563–588.

Riley, J. (1989). *Sickness, recovery and death: A history and forecast of ill health*. Iowa City: University of Iowa Press.

Riley, J. (1990). The risk of being sick: Morbidity trends in four countries. *Population and Development Review, 16*, 205–232.

Riley, J. (1991). The sick and the well: Adult health in Britain during the health transition. *Health Transition Review, 6*(suppl), 19–44.

Riley, J. (1997). *Sick, not dead. The health of British workingmen during the mortality decline*. Baltimore: Johns Hopkins University Press.

Riley, J. (1999). Why sickness and death rates do not move parallel to one another over time. *Social History of Medicine, 12*, 101–124.

Riley, J., & Alter, G. (1989). The epidemiologic transition and morbidity. *Annales de Demographie Historique*, 199–213.

Robine, J. M., Jagger, C., Mathers, C. D., Crimmins, E. M., & Suzman, R. M. (2003). Trends in health expectancies. In J. M. Robine, C. Jaggers, C. D. Mathers, E. M. Crimmins, & R. M. Suzman (Eds.), *Determining health expectancies*. Chichester: Wiley.

Romer, P. M. (1990). Endogenous technological change. *Journal of Political Economy, 98*(5), S71–S102.

Sen, A. K. (1998). Mortality as an indicator of economic success or failure. *The Economic Journal, 108*(446), 1–25.

Steckel, R. H., & Costa, D. L. (1997). Long-term trends in health, welfare and economic growth in the United States. In R. Floud & R. H. Steckel (Eds.), *Health, welfare and industrialization*. Chicago: University of Chicago Press.

Thatcher, A. (1992). Trends in numbers and mortality at high ages in England and Wales. *Population Studies, 46*, 411–426.

Thatcher, A. (1997). Trends and prospects at very high ages. In J. Charlton & M. Murphy (Eds.), *The health of adult Britain 1841–1994* (Vol. 2). London: Office of National Statistics.

Verbrugge, L. M. (1984). Longer life but worsening health? Trends in health and mortality of middle-aged and older persons. *Milbank Memorial Fund Quarterly, 62*, 475–519.

Weisbrod, B. A. (1991). The healthcare quadrilemma: An essay on technological change, insurance, quality of care, and cost containment. *Journal of Economic Literature, 29*, 523–552.

Appendix

The history of the modern International Classification of Diseases (ICD) can be traced to William Farr who devised in 1839 a system of disease classification, which was revised later in the 1850s and 1860s. It had five main groups: (1) epidemic; (2) constitutional (general); (3) local (by anatomical site); (4) developmental; (5) violence. (For a detailed account in England and Wales see Eyler 1979; Hardy 1994; Rob-Smith 1969, 1970). Farr's classification largely became the basis for the International Classifications in the twentieth century. Though the international lists have been revised essentially every decade in the twentieth century, the principle of classifying disease largely by anatomical site (and to some extent by etiology) has been maintained through all revisions. The original list was proposed for international consideration in 1855 (though variations of it had been used earlier in England). It was then revised in 1881 and in 1891. The first International Classification was issued in 1893 by the International Statistical Congress; the Bertillon list or ICD-1 was issued in 1901 and was adopted/adapted by many nations. The subsequent lists were issued by the League of Nations, and after World War II, by the World Health Organization. The ICD revisions are: ICD-1 (1901–1910), ICD-2 (1911–1920), ICD-3 (1921–1930), ICD-4 (1931–1939), ICD-5 (1940–1949), ICD-6 (1850–1857), ICD-7 (1958–1967), ICD-8 (1968–1978), and ICD-9 (1979–2000).

To rebuild the time series of various disease categories across the classifications, scholars usually adopt a recent ICD list and track its constituents though time (e.g. Charlton and Murphy 1997). Following that cue, the book adopts the ninth revision of the International Classifications (ICD-9, WHO 1978) and isolates since the mid-nineteenth century the two main disease categories discussed in the book. The composition of *infectious diseases* is listed in Tables A.1 and A.2. For infectious diseases in ICD-1 and for the classifications before it, Table A.2 lists the diseases by name (instead of the codes) since they did not have an international code. In the Table, asterisks mark the diseases that had been originally classified under a different rubric and had to be brought in to make the composition as consistent as possible over time. The main reason for separating out this infectious

S. Arora, *The Transitions of Aging*, International Perspectives on Aging 12,
DOI 10.1007/978-3-319-14403-0

sub-category is that they used to be readily identifiable on clinical grounds and physicians are likely to have been familiar with such diseases (Williams 1996; McKeown 1976). Nonetheless, in the annual reports, (1) many individual infectious diseases were combined with others and cannot be isolated; (2) the reportage of many has become more granular over time; (3) many infectious diseases can have similar symptoms and can be easily confounded. For all three reasons, it is safer, for the purpose of this study, to track them as a broad aggregate.

Likewise, Table A.3 lists all violent or external causes such as injuries, accidents and poisonings (referred as 'external causes' in the text). Table A.4 lists the pregnancy complications.

As mentioned in Chap. 2, the broad category 'non-communicable disease' was calculated as: All causes minus all infectious diseases minus all 'external causes' minus complications of pregnancy and childbirth.

The Registrar General's Annual Reviews since the mid-nineteenth century (and later, Office of National Statistics), and the World Health Statistics are the main sources. The series were built up from annual age-specific raw data. The age-specific data where converted to crude rates. The crude rates were then standardized based on 1994 weights where necessary (e.g. for the aggregate-level trends in Chaps. 2, 3, and 4). The data on annual population by age are from the Human Mortality Database.

The use of broad sub-categories, as opposed to individual diseases within each category, is safer for cross-decade comparisons. In each broad sub-category, the codes for individual diseases have become increasingly granular over time and many individual diseases are difficult, if not impossible, to reconstruct consistently across the revisions. Though the compositions of the broader categories, too, have been revised, the errors in their reconstruction are likelier to be less as the bulk of the revisions have pertained to within-category changes.

For that reason, the aggregate non-communicable diseases, as defined, produce the most reliable results, especially over the adult-age segment of the life-cycle. Over the years, apart from minor cases such as aortic syphilis (transferred from overall syphilis as an infectious disease to the sub-category circulatory system) or non-epidemic meningitis (transferred to nervous system), the composition of non-communicable diseases as an aggregate is more consistent than the composition of any of its sub-categories because any transfers between its sub-categories still remains within the aggregate 'non-communicable disease.' The confidence in any series, however, should diminish as one moves from that very broad level of aggregation to the sub-category-level. Among the Tables below, I have the least confidence in Neoplasms, especially before the 1890s and to a large extent in the nervous system diseases as many of them had been classified in the ICD lists before the 1890s as circulatory system diseases and the details reported may not have been enough to cleanly transfer from one-category to the other.

Besides differences across the lists, the varying coverage of registration may influence the data over time. In the very early years of death registration, certification is likelier to have been more inaccurate than today and some deaths may not have been registered at all (Wrigley and Schofield 1981). The onus to register used to be on the registrar's office instead of the informant. However, the informant could be prosecuted on refusal to comply with the registrar's query. All told, incomplete death registration was less of an issue as registration was legally required for burial. In the 1850s, 1860s and 1870s, the vast majority of deaths were reported by qualified practitioners familiar with the lists issued by the GRO (Williams 1996; Alter and Carmichael 1996). An 1874 law then made medical practitioners responsible for the information; the non-compliant were subject to penalty. In 1878 only about 5% of the deaths were not certified by a medical practitioner; By 1891–1900, about 2.3% not certified; by 1928, about 1% (Ashley and Devis 1992).

In 1927, the format of the death certificate changed, requiring a primary as well as a secondary cause (or accompanying causes); the certificate remained similar through 1990s. According to WHO (1978), the primary or the underlying cause is: (1) the disease or injury that initiated the chain of events leading to death; or (2) the circumstances of the accident or violence (e.g. suicide) that produced the fatal injury. Prior to 1940, however, the selection by the certifying practitioner was rule-based, whereas afterward it was in accordance of the certifier's preference expressed in the order on the certificate (Campbell 1965; Logan 1950), though the change did not seem to have altered the long-term trends since the nineteenth century. To gauge within-sub-category accuracy, Heasman and Lipworth (1966) studied 9,501 patients in 75 hospitals in England and Wales, and found disagreements in only 45% of the cases, but the overall numbers within the sub-categories was similar to the originals as the errors tended to cancel out. Alderson (1981) reviewed a number of validation studies to assess accuracy of the cause of death certification and found that in general 70% of the cases were accurate the first time over, with minor revisions required in 20% of the cases and only in 10% of cases was major change required.

All told, it appears that a safer way to use the data over the long-term, across ICD revisions, is at much broader levels of aggregation than at the individual disease level (Tables A.5, A.6, A.7, A.8, and A.9).

Table A.1 Infectious disease (as defined; see text, Chap. 2), ICD-2 through ICD-9

ICD-9	1–136; 460–519
ICD-8	1–136; 460–519
ICD-7	1–138, 571, 240–41, 470–527
ICD-6	1–138, 571, 240–41, 470–527
ICD-5	1–32, 34–44, 81, 119–120, 177; 33, 10–114, 115 (c)
ICD-4	1–10, 12–44, 79–80, 83, 119–120, 177, 11, 104–114, 115(2–4)
ICD-3	1–10, 12–42, 71, 72, 76, 113–116, 121, 175, 11, 109, 97–107
ICD-2	1–9, 11–25, 28–35, 37–38, 60–62, 67, 104–107, 112, 164, 10, 86–98, 100

Table A.2 Infectious disease ICD-1 and older lists

ICD-1(1901)	1881–1900	1861–1880	1855–1860
Smallpox	Smallpox	Smallpox	Smallpox
Cowpox	Chickenpox	Chickenpox	Chickenpox
Chickenpox	Measles	Measles	Miliaria
Measles (Morbilli)	Epidemic rose rash	Scarlet fever (Scarlatina)	Measles
German measles	Scarlet fever	Diphtheria	Scarlatina
Scarlet fever	Typhus	Quinsy	Cynache Maligna
Typhus	Relapsing fever	Croup	Diphtheria
Plague (all types)	Whooping cough	Whooping cough	Mumps
Relapsing fever	Mumps	Typhus	Whooping cough
Whooping cough	Diphtheria	Erysipelas	Croup
Mumps	Cerebrospinal fever	Carbuncle	Thrush
Diphtheria	Simple, ill-defined fever	Influenza	Typhus
Cerebrospinal fever	Enteric fever	Dysentery*	Diarrhea*
Pyrexia (uncertain)	Other miasmatic	Diarrhea*	Dysentery*
Enteric fever	Cholera	Enteric fever	Cholera
Asiatic cholera	Diarrhea*	Cholera	Influenza*
Diarrhea (food-caused)	Dysentery*	Ague	Ague
Infective enteritis*	Remittent fever	Remittent fever	Remittent fever
Epidemic diarrhea*	Ague	Malaria	Yellow fever
Dysentery*	Hydrophobia	Mumps	Typhus
Tetanus*	Glanders	Erythema	Erysipelas
Malaria (all types)*	Anthrax, Splenic fever	Yellow fever	Phlebitis
Rabies, Hydrophobia	Cowpox	Syphilis	Malignant pustule
Glanders	Syphilis	Gonorrhea, Stricture urethra	Glanders
Anthrax	Gonorrhea, Stricture urethra	Hydrophobia	Porrigo

(continued)

Table A.2 (continued)

ICD-1(1901)	1881–1900	1861–1880	1855–1860
Syphilis	Phagadena	Glanders	Syphilis
Gonorrhea	Erysipelas	Thrush	Hydrophobia
Erysipelas	Pyaemia	Worms	Scrofula
Septicemia (non-puerperal)	Septicemia (non-puerperal)	Porrigo*	Tabes mesenterica
Pyaemia	Thrush*	Tapeworms*	Tubercular peritonitis
Phlegmon, carbuncle*	Other veg. parasites*	Hydatids*	Phthisis
Phagadena	Hydatid disease*	Ascarides Lumbricoides*	Hydrocephalus*
Other infective processes	Animal parasites*	Scrofula	Tetanus*
Pulmonary tuberculosis	Tabes Mesenterica	Tabes Mesenterica	Hydrophobia
Tuberculous meningitis	Phthisis (lung TB)	Tubercular Peritonitis	Syphilis
Tuberculous peritonitis	Other TB, Scrofula	Phthisis (lung TB)	Cephalitis
Tabes mesenterica	Inflammation of brain*	Hydrocephalus	Gastritis*
Tubercle, other organs	General paralysis of insane	Cephalitis	Enteritis*
General tuberculosis	Idiopathic tetanus	Idiopathic tetanus*	Hepatitis*
Scrofula	Enteritis*	General paralysis of insane	Jaundice*
Parasitic diseases*	Gastro-enteritis*	Enteritis*	Carbuncle*
Meningitis, inflammation of brain*	Carbuncle*	Hepatitis*	Laryngitis*
General paralysis of insane	Croup	Jaundice*	Emphysema
Locomotor ataxy*	Food poisoning	Laryngitis*	Bronchitis
Croup	Influenza*	Emphysema	Pleurisy
Influenza*	Laryngitis	Bronchitis	Pneumonia
Laryngitis	Diseases of Larynx and Trachea	Pleurisy	Asthma
Other diseases larynx	Bronchitis	Pneumonia	Other lung diseases
Bronchitis	Emphysema, Asthma	Asthma	Pharyngitis
Emphysema, Asthma	Pleurisy	Influenza	
Pleurisy	Pneumonia	Other diseases of lungs	
Fibroid disease of lung	Other respiratory	Pharyngitis	
Other respiratory diseases	Quinsy		
Tonsilitis			
Quinsy			
Disease of the pharynx			
Pneumonia			

Table A.3 Complications of pregnancy, childbirth and puerperium

ICD-9	630–679
ICD-8	630–679
ICD-7	640–689
ICD-6	640–689
ICD-5	401–503
ICD-4	400–503
ICD-3	431–500
ICD-2	134–141
ICD-1 and *before*	Puerperal Septicaemia, Septic intoxication, Puerperal Pyaemia, Phlegmasia Alba Dolens, Puerperal fever (not defined), Abortion, Miscarriage, Puerperal mania and convulsions, Placenta Praevia, Flooding, Other complications; (Metria (including Puerperal mania and Convulsions), Childbirth, Puerperal fever, Paramenia (less Chlorosis)

Table A.4 Injury and poisoning ('external causes')

ICD-9	E800-999
ICD-8	E800-E999
ICD-7	E800-999
ICD-6	E800-999
ICD-5	163–176, 178–198
ICD-4	163–175, 178–198
ICD-3	165–174, 176–203
ICD-2	57–58, 153, 155–163, 165–173, 174–186
ICD-1 and *before*	Violent deaths: in mines and quarries; vehicles and horses, building operations, machinery, weapons and implements, burns and scalds, poisons, drowning, suffocation, falls, weather agencies, battle, homicide, suicide, execution. (Accidents or negligence (fractures and contusions, gunshot wounds, cut, stab, burns and scalds, poison, drowning, suffocation, others, murder and man-slaughter, suicide, execution, other violent deaths)

Table A.5 Circulatory system

ICD-9	390–459
ICD-8	390–444, 444–589, 782–89
ICD-7	330–334, 400–468 (less 455), 782
ICD-6	330–334, 400–468 (less 455), 782
ICD-5	58, 83 (less 83e), 87a, 90–97, 99–103
ICD-4	56. 82, 87a, 90–97, 99–103
ICD-3	51, 74, 81, 83, 87–96
ICD-2	47, 64–65, 72, 77–85
ICD-1 and *before*	Acute articular rheumatism, Rheumatic fever*, cerebral hemorrhage and apoplexy, Softening of brain, Chorea*, Pericarditis, Acute Endocarditis, Valvular Disease, Hypertrophy of heart, Dilation of heart, angina pectoris, Fatty degeneration of heart, Aneurysm, Non-cerebral Embolism and Thrombosis, Phelbitis, Varicose veins, Other diseases of blood vessels, Other diseases of the heart, Syncope (unspecified heart disease), Diseases of the lymphatic system, Hemorrhage

Table A.6 Nervous system and organs of special sense

ICD-9	320–459
ICD-8	320–389, 733–781
ICD-7	335–398, 740–744
ICD-6	335–398, 740–744
ICD-5	80–82, 83e, 85, 87b, 87c, 87d, 88, 89
ICD-4	78-79, 81, 85, 87b, 87c, 87d, 87e, 88, 89
ICD-3	70, 73, 75, 78, 79–80, 82, 84(3), 84(4), 84(5), 85, 86
ICD-2	60, 61b, 61c, 63, 66, 69, 73, 74a, 74b, 74d, 75, 76
ICD-1 and *before*	Epilepsy, Periplegia, Diseases of Spinal Chord, Peripheral Neuritis, Polyneuritis, Other diseases of the nervous system, Otitis (otorrhoea), Mastoid disease, Opthalmia, Diseases of eyes (including cataract, etc.), Epistaxis and other diseases of the nose, Other disease of the brain; (Nercrencephalus, Spinal Marrow disease, Neuralgia, Shaking palsy, Undistinguished brain disease)

Table A.7 Digestive system

ICD-9	520–577
ICD-8	520–577
ICD-7	530–570, 572–587
ICD-6	530–570, 572–587
ICD-5	115b, 116–118, 121–129
ICD-4	115a, 115b, 116–118, 121–129
ICD-3	108, 110–112, 117–120, 122–127
ICD-2	99, 101–103, 108–111, 113–115, 117–118
ICD-1 and *before*	Gastric Ulcer, Gastric Catarrh, Other diseases of the stomach, Appendicitis, Perityphilitis, Hernia, Intestinal Obstruction, Other diseases of the intestines, Peritonitis, Cirrhosis of liver, Other diseases of liver and gall bladder, Other diseases of the digestive system; (Ascites, ulceration of intestines, Ileus, Intussusception, Stricture of intestinal canal, Diseases of stomach, Diseases of liver)

Table A.8 Musculoskeletal system

ICD-9	710–739
ICD-8	710–738
ICD-7	710–732, 734–738
ICD-6	720–743, 745–749
ICD-5	58b, 59, 154–156
ICD-4	57, 154–156
ICD-3	52, 155–158
ICD-2	48, 146–149
ICD-1 and *before*	Chronic Rheumatism, Rheumatic Arthritis, Rheumatic Gout, Gout, Caries, Necrosis, Arthritis, Periostitis, Other diseases of locomotion; (Orthritis, rheumatism, Diseases of joints, Gout, Diseases of the organs of locomotion)

Table A.9 Neoplasms

ICD-9	140–239
ICD-8	140–239
ICD-7	40–239, 294
ICD-6	40–239, 294
ICD-5	44–57, 74
ICD-4	44–55, 72
ICD-3	43–49, 50, 65, 84b, 139
ICD-2	39–45, 46, 74c, 53, 129
ICD-1 and *before*	Carcinoma, Sarcoma, Cancer and Malignant Disease, Anaemia and Leucocythaemia, Brain tumor, Ovarian tumor, Tumor, Uterine Tumor; (Cancer, Melanosis, Sweep's Cancer, Lupus, Polypus, Undistinguished Cancer, Cancrum Oris (Noma), Abcess, Tumor)

References

Alderson, M. R. (1981). *International mortality statistics*. London: McMillan.

Alter, G., & Charmichael, A. (1996, Spring). Studying causes of death in the past: Problems and models. *Historical* Methods, *29*(2), 44–48.

Ashley, J., & Devis, T. (1992). Death certification from the point of view of the epidemiologist. *Population Trends, 67*, 22–28.

Campbell, H. (1965). *Changes in mortality trends: England and Wales 1931–1961* (National Center for Health Statistics, Series 3(no. 3) US DHEW). Washington, DC: Public Health Service.

Eyler, J. (1979). *Victorian social medicine: The ideas and methods of William Farr*. Baltimore: Johns Hopkins University Press.

Hardy, A. (1994). 'Death is a cure of all disease:' Using the General Register Office cause of death statistics for 1837–1920. *Social History of Medicine, 7*(3), 472–492.

Heasman, M. A., & Lipworth, L. (1966). *Accuracy of certificates of causes of death* (OPCS Studies in Medical and Population subjects no. 20). London: HMSO.

HMD (Human Mortality Database), University of California at Berkeley & Max Plank Institute for Demographic Research. www.mortality.org

HMSO. (1857). *The nomenclature of diseases drawn up by the Joint Committee appointed by the Royal College of Physicians of London* (1st, 2nd, 3rd ed.). London: HMSO.

HMSO/ONS. *Registrar general's statistical review of England and Wales, various years 1850–2000*. London

Logan, W. P. D. (1950). Mortality in England and Wales from 1848–1947. *Population Studies, 4*, 132–178.

Mckeown, T. (1976). *The modern rise of population*. London: Edward Arnold.

ONS. (1997). *Health of adult Britain 1841–1994*, Vol. I and II (Decennial Supplement no. 12 and 13), edited by J. Charlton & M. Murphy. London.

Rob-Smith, A. H. T. (1969). A history of the college's nomenclature of diseases. *Journal of the Royal College of Physicians of London, 3*, 341–358.

Rob-Smith, A. H. T. (1970). A history of the college's nomenclature of diseases. *Journal of the Royal College of Physicians of London, 4*, 2–26.

Williams, N. (1996, Spring). The reporting and classification of causes of death in mid-nineteenth century England. *Historical Methods, 29*(2), 58-70.

World Health Organization. (1978). *Manual of international statistical classification of diseases, injuries and causes of death: Ninth revision*. Geneva: World Health Organization.

Wrigley, E. A., & Schofield, R. S. (1981). *The population history of England, 1541–1871: A reconstruction*. Cambridge, MA: Harvard University Press.

Index

S. Arora, *The Transitions of Aging*, International Perspectives on Aging 12,
DOI 10.1007/978-3-319-14403-0